GeoMeasurements
by Pulsing TDR
Cables *and* Probes

GeoMeasurements
by Pulsing TDR
Cables *and* Probes

Kevin M. O'Connor, Ph.D., P.E.
President
GeoTDR, Inc.

Charles H. Dowding, Ph.D., P.E.
Professor of Civil Engineering
McCormick School of Engineering
and Applied Science
Northwestern University

CRC Press
Taylor & Francis Group
Boca Raton London New York

CRC Press is an imprint of the
Taylor & Francis Group, an **informa** business

CRC Press
Taylor & Francis Group
6000 Broken Sound Parkway NW, Suite 300
Boca Raton, FL 33487-2742

First issued in paperback 2019

© 1999 by Taylor & Francis Group, LLC
CRC Press is an imprint of Taylor & Francis Group, an Informa business

No claim to original U.S. Government works

ISBN-13: 978-0-8493-0586-3 (hbk)
ISBN-13: 978-0-367-39997-9 (pbk)

Library of Congress Cataloging-in-Publication Data

O'Connor, Kevin M.

    Geomeasurements by pulsing TDR cables and probes / Kevin M.
O'Connor, Charles H. Dowding
       p. cm.
    Includes bibliographical references and index.
    ISBN 0-8493-0586-1 (alk. paper)
    1. Engineering geology--Instruments. 2. Time-domain
reflectometry. 3. Soils--Testing. 4. Rocks--Testing. I. Dowding,
Charles M. II. Title.
TA705.028 1999
624.1'51'028—dc21

Library of Congress Card Number 98-40834

98-40834
CIP

**Visit the Taylor & Francis Web site at
http://www.taylorandfrancis.com**

**and the CRC Press Web site at
http://www.crcpress.com**

Preface

The authors are indebted to many individuals and organizations who have contributed to our collective experience. Their support was critical during this more than five-year-long project to unify the field of TDR instrumentation. Two organizations stand out during this process because of their long-term support: the U.S. Bureau of Mines and the Infrastructure Technology Institute at Northwestern University.

After the pioneering work by Dowding, O'Connor, and Su (1989) that substantiated the ability to measure deformation in rock with TDR cable deformation, the U.S. Bureau of Mines became the principal focus for development of TDR technology to monitor deformation and water levels. The Bureau served as Dr. O'Connor's home base and supported telecommunication work at Northwestern.

The Infrastructure Technology Institute has supported development of a miniaturized, low-power-consumption TDR pulsing board and the commercialization of the technology for use in monitoring critical highway facilities. This support provided the funds that allowed the production of this book.

Support of two other organizations has also been helpful in this project: the U.S. Army Corps of Engineers (ACE) and the National Science Foundation. The ACE adopted development of the miniaturized pulser for water level measurement as a Construction Productivity Advancement Research (CPAR) project. They were particularly interested in retrofitting piezometers that measure water pressure beneath dams with hollow coaxial cables to telemetrically monitor dam stability. The NSF has begun to sponsor research to develop a grout and cable system that is sufficiently compliant to monitor shearing of soft soils with TDR.

Special thanks is due to the organizing committee of, and key contributors to, the 1994 First International TDR Symposium. Particularly helpful were Dr. William N. Herkelrath of the U.S. Geological Survey, Dr. Joel Greene of Campbell Scientific Inc., Dr. Clark Topp of Agriculture Canada, Dr. Alan D. Kersey of the

Naval Research Laboratory, Dr. Dryver Huston of the University of Vermont, Dr. James W. Ogle and Dr. Ed Van Eeckhout of Los Alamos National Laboratory, Dr. Michael Norland of the U.S. Bureau of Mines, Dr. James Andrews of Picosecond Pulse Labs Inc., Dr. R. Gary Kachanoski of the University of Guelph, Mr. Curtis Smith of Tektronix Inc., Drs. Steve Zegelin and Ian White of CSIRO, and Dr. Lewis V. Wade of the U. S. Bureau of Mines. The proceedings for the symposium provided a great deal of the material for this book. We are grateful for all of information contributed by authors of the papers presented in those proceedings, who are too numerous to mention. Their contributions have been cited throughout this book.

Preliminary drafts of each chapter were reviewed by individuals with expert knowledge of the various applications of TDR technology: Dr. Clark Topp and Dr. Steve Zegelin (Chapters 2 to 4), Dr. Cathy Aimone-Martin and Dr. William Kane (Chapters 5 to 7), Dr. Dryver Huston (Chapter 8), Mr. Douglas Bailey (Chapter 9), and Mr. Russell Anderson (Chapters 10 and 11). Copy editing was done by Mr. Michael McNett and Ms. Mimi Williams.

Recognition is also extended to personnel at Northwestern who contributed to this project. We are grateful to all of the graduate students who through the years have contributed to advancements in this field through their theses and other assistance: N. Amyrd, F. Blais, C. Kroll, A. Graettinger, P. McComb, Dr. F. C. Huang, Dr. C. E. Pierce, Dr. M. B. Su, D. Will, and Dr. G. Williams. We are also grateful to the support staff who contributed to the production of this book: Sabrina Goodman, Renee McHenry, and Brian Nielsen. Special thanks to Joel Meyer whose artists drew the illustrations.

Many of the practical examples have resulted from consulting assignments and cooperative research agreements with groups such as the Illinois Mine Subsidence Insurance Fund, Peabody Coal Co., Ohio Department of Transportation, California Department of Transportation (CALTRANS), Syncrude Canada Ltd., PacifiCorp, Mettiki Coal Co., Old Ben Coal Co., Western-AG Minerals, WIPP, Pennzoil Sulphur Co., CANMET-MRL, Island Creek Coal Co., Freeman United Coal Co., Collinsville Community Unified School District 10, Illinois State Geologic Survey, U.S. Geological Survey,

HYPERLABS, Campbell Scientific Inc., ESI Environmental Sensors, Inc., and PermAlert. The experience from this interaction has proven invaluable in providing the practical, field perspectives throughout this book.

We thank everyone who helped, both named and unnamed. We are especially indebted to our wives, Jill O'Connor and Jane Dowding, and to our children who allowed the book preparation time to be subtracted from our otherwise busy lives and devoted to this project.

ABOUT THE
AUTHORS

KEVIN M. O'CONNOR

Kevin M. O'Connor is President of GeoTDR, Inc., Apple Valley, Minnesota. He obtained a Ph.D. and M.S. in Civil Engineering from Northwestern University, and a B.S. in Civil Engineering and B.S. in Geology from the University of Notre Dame. He is a registered professional engineer and a licensed geotechnical engineer who has been involved with developing applications of TDR technology for 17 years. While at the U.S. Bureau of Mines in Minneapolis, MN, he was the principal investigator responsible for development, installation, and analysis of TDR monitoring systems in a variety of applications throughout the U.S. and Canada. He was coorganizer and proceedings editor of the Symposium on Time Domain Reflectometry in Environmental, Infrastructure, and Mining Applications held at Northwestern University in September, 1994. Following closure of the USBM, he formed GeoTDR, Inc. to continue application and development of TDR technology. In addition to presenting short courses, he has authored or coauthored over fifteen technical publications on TDR technology.

CHARLES H. DOWDING

Charles H. Dowding is Professor of Civil Engineering at Northwestern University in Evanston, Illinois. He obtained a Ph.D. and M.S. in Civil Engineering from the University of Illinois at Urbana and his B.S. in Civil Engineering from The University of Colorado at Boulder. He is a registered professional engineer who has been involved with developing applications of TDR technology for 15 years. He is the originator of the concepts for using TDR to monitor and quantify deformation in rock and to monitor changes in ground water levels. He was coorganizer and chairman of the

TDR symposium and continuously explores new applications of TDR technology. He has authored or coauthored over fifteen technical publications on the development and application of TDR technology as well as two other books, _Construction Vibrations_ and _Blast Vibration Monitoring and Control_.

Contents

APPENDICES

Chapter 1
INTRODUCTION

This book project was written to crystallize the commonalities among the seemingly divergent specialties employing Time Domain Reflectometry (TDR) technology in geomaterials. The diversity of interest and terminology within the TDR community tends to obscure commonalities and the fact that physical principles underlying the technology are universal. It is our hope that the condensation, distillation, and integration of the varied TDR experiences in this book will provide the synergism necessary to unify the field.

EVOLUTION OF TDR TECHNOLOGY

TDR technology was first developed during the 1950s to locate and identify cable faults in the power and telecommunications industries which remain major users of this technology. TDR cable testers are now considered standard equipment by engineers and technicians in these major industries. The enormous size of the power, telecommunications, and more recently the cable TV and computer network industries provides the largest market for the development of TDR hardware.

In the 1970s, TDR technology began to be applied with geomaterials, and it is this expansion of use and technology transfer that led to the writing of this book. Now TDR technology is employed by soil scientists, agricultural engineers, geotechnical engineers, and environmental scientists as well as electrical engineers. Initial empirical methods of application in geomaterials have matured with scientific calibration and improved pulser samplers, cables, and sensors. These accelerating improvements are condensed here.

This book represents another step along the evolution of TDR technology in geomaterials from a research instrument to a robust, reliable, and economical production tool. Thus far, users of TDR in geomaterials typically have been researchers and this book attempts to clear the way for use of TDR by a variety of non-specialists. A major effort has been made to incorporate experience and reports of field verification, and their inclusion not only substantiates TDR's usefulness, but also provides comparisons with other commonly employed technologies. The broad range of positive experience supports the assertion that TDR technology is not a passing fad but has already stood the test of time.

Shelves of articles as well as research and consulting documents have been condensed into this book to provide the perspective on application of TDR technology that is not achievable one article at a time. Basically, this book expands upon the First International Symposium on Time Domain Reflectometry in Environmental, Infrastructure, and Mining Applications that was held at Northwestern University in 1994. That symposium brought together for the first time the incredibly diverse disciplines employing TDR and allowed experts to exchange experiences. Even though these experiences were exchanged through the proceedings of the symposium, as with all such proceedings, they were not integrated. To the base of experience provided by the 1994 symposium, the authors have added their own. This additional experience has focused on many of the secondary applications rather than the currently most active one, soil moisture monitoring.

TDR'S BROAD RANGE OF APPLICATION

In general, Time Domain Reflectometry describes a broad range of remote sensing electrical measurements to determine the location and nature of various reflectors. TDR is similar in principle to radar which consists of a radio frequency transmitter (which emits a short pulse of electromagnetic energy), a directional antenna, and sensitive radio frequency receiver. After the transmitter has directionally radiated the pulse, the receiver records the echo or reflection returned from a distant object such as an airplane or ship. By measuring the time between transmission and receipt of the reflection and knowing the speed of light, the distance to the object may be easily calculated. Detailed analyses of the echo can reveal additional details of the reflecting object, which aid in identification.

Essentially TDR is radar along a coaxial cable that can enable surveillance of large volumes with a single instrument. Distances to reflectors along the cable are calculated by time of flight, and characteristics of reflectors anywhere along a cable can be discerned from details of their respective reflected signals. The ability to interpret TDR reflections anywhere along the cable allows activity to be monitored over large volumes or areas and thus TDR monitoring may replace many single-point measurement instruments. This inherent surveillance advantage is propelling new applications of TDR in geomaterials for monitoring chemical spills, oil leaks beneath storage tanks, movement of rock in mines, and potential slope failures in soft soils.

To date, TDR's dominant application in geomaterials has been the measurement of moisture (or water content) of unsaturated soils. This application occurs at a probe at the end of the cable or two-wire transmission line. Thus, unlike "along the cable" applications, the probe locations must be chosen at the time of placement. Moisture (water content) measurements are made in soil for irrigation research and control, soil covers for landfills, base courses beneath highway pavements, bulk storage piles of minerals, and other granular materials where water content is important.

Detection of fluids is an emerging TDR application. Liquid interfaces produce significant reflections in hollow and porous cables, which can be easily detected. This observation has led to the use of TDR for leak and pollution detection along a selectively porous cable, which allows monitoring of large volumes with a single cable. In addition, fluid level detection techniques have led to use of TDR to measure water levels for hydrological purposes, as well as for measurement of water pressures beneath dams.

WHAT IS THE SIZE OF THE TDR MARKET ?

By far the largest markets for TDR technology involve hundreds of thousands of units in the telecommunications and power industries. Thus equipment manufacturers have focused on these markets. Because of the large size of this communications market, many of the major companies regard the geomaterials and infrastructure applications discussed in this book as niche markets. At present, use of TDR in geomaterials is dominated by the study, monitoring, and control of soil water content for conservation of agricultural water. Since the potential number of soil moisture or irrigation control systems is measured in terms of a hundred thousand worldwide, it is no wonder that several firms have been formed to serve this potential market.

There are a number of applications of TDR technology that are rendered potentially commercially viable by the infrastructure provided by the level of activity in the areas of telecommunications and soil moisture. In order of decreasing numbers of potential installations they are:

> leak detection and contaminant monitoring
> water level measurement
> > hydrological and contaminant transport
> > water pressures
> deformation and stability monitoring of
> > mines (abandoned and operating)
> > slopes (rock and soil)
> > structures (bridge scour, earthquake degradation)

The leak detection market alone is estimated to be in the hundreds of thousands in the United States with the requirement of monitoring systems for all petroleum storage facilities over one thousand gallons. Added to that is the potential to employ TDR technology for the detection of other chemical and polluting liquids. The water level market is estimated to be in the tens of thousands and is also important for monitoring safety of water diversion structures. The market for deformation monitoring is estimated to be in the thousands and could be larger if and when automated telecommunication systems become robust enough to be installed and maintained by local personnel.

Broadening the market for TDR technology is critical for generating the enthusiasm needed to ensure the continued development of necessary components such as miniaturized pulsers and specialty cables and probes. Our interaction with original equipment

manufacturers (OEMs) and venture capitalists has indicated that
markets of 100s to 1,000s hold no promise for investment, no
matter how helpful the technology. Markets of 1,000s to 10,000s
attract attention but little investment from OEMs (at this level
forgivable, supplementary funding from Federal sources is necessary
to attract OEM participation). Only at the 10,000 to 100,000 unit
threshold is there significant interest. It seems that the markets for
leak detection and soil moisture measurement, which when
combined could approach multiple hundreds of thousands, justify
the continued development of TDR instruments as well as specialty
cables and transducers.

WHO SHOULD READ THIS BOOK

Because of the broad spectrum of measurement possible with TDR,
this book should be of interest to a wide variety of scientists,
engineers, and contractors. Consider the following list of disciplines
and applications of TDR technology:

> soil scientists and agricultural engineers monitor soil mois-
> ture,
> mining engineers monitor rock mass deformation,
> chemical engineers and facility managers monitor chemical
> and petroleum spills,
> civil engineers monitor water content of landfill covers and
> water pressures beneath dams, and
> geotechnical engineers monitor soil slope movements.

While the greatest number of installations are employed to monitor
soil moisture (water content) and chemical and petroleum spills, use
of TDR to monitor deformation is growing. Needs of all these
disciplines are addressed in this book.

Material is presented at a professional level; it is assumed
that the reader is college educated. Thus the book is appropriate
for professional practice and graduate level courses in engineering
and soil science. As far as possible, the presentation follows an
observational or empirical rather than mathematical approach.

However, the basic physics and quantitative nature of signal interpretation and computer processing will of necessity require some mathematics.

FUTURE DEVELOPMENTS

This chapter began with a short history, and the introductory process can be completed by concluding with an estimate of future developments. In addition to symmetry there are several important reasons to peer over the horizon. First, assessing future developments places in perspective the accomplishments chronicled herein. Second, contents of all books lag the forefront of research because they must be frozen for editing and publication at least a year before they are available for sale. Thus in a sense there are two time frames relative to this book. Near-term developments which are already well underway and long-term developments which are conceptual. Perhaps most important, TDR is an emerging technology for use in geomeasurements which is on the accelerating portion of the development curve. Its potential should not be judged solely by the snapshot in time represented by this book.

So what does the future hold? It will definitely involve many more users. For instance in the last year, the authors have installed cables to monitor deformation in unusual geo-materials for a number of new clients. Kilometers of leak detection cables and hundreds of moisture sensors are being installed at new sites each year. These clients in turn will use it on more projects and development of new products will accelerate.

Continued miniaturization of digital telemetry makes remote monitoring a practical reality. The inherent digital nature of TDR measurements will further accelerate its use for remote monitoring. Already, GeoTDR and Northwestern University have been involved with remotely operable systems at more than a dozen sites world-wide.

Smaller, less expensive pulser-samplers are being developed. For instance, the HYPERLABS instrument has already been modified and is being marketed. Another TDR unit has been developed as a plug-in card for use with portable field computers.

There will be new specialty cables and new uses for existing cables. Work has begun at Northwestern University to build a large-diameter compliant cable for use in soft soils. Prototypes could be installed within two years. Air-dielectric cables are being used to retrofit existing piezometers as described in Chapter 9. Deformation-monitoring cables are being placed in obstructed inclinometer casing to extend the useful life of these installations.

There will be new probes and calibration techniques for measurement of soil moisture. Inexpensive methods are being developed to field calibrate moisture probes for specific soils. TDR is emerging as a viable alternative to nuclear techniques for measurement of *in situ* density of soils.

Look for greater use of TDR probes and cables to monitor contaminant migration. Already, air-dielectric cables are being employed by GeoTDR to monitor changes in subsurface oil pools. In this application, the difference of dielectric properties of oil and underlying ground water allows the thickness of oil to be measured directly. Companies are continuing to develop new specialty cables for leak detection and more cables will be employed to retrofit existing installations.

Finally, new analytical techniques will be developed. As described above, soil moisture probes are being calibrated for specific soils. Likewise, models are being developed that allow more sophisticated analysis of TDR signatures generated by cable deformity. For example, work will be completed this year at Northwestern on a numerical model that incorporates frequency dependent signal loss with transmission distance. This model will allow calibration of cables as well as analysis of multiple reflections.

INTERNET SUPPORT

This book is supported extensively on the Internet. The Infrastructure Technology Institute at Northwestern University maintains a TDR clearinghouse that contains full text copies of articles, addresses of vendors, and links to related technology. It can be reached at

http://www.iti.nwu.edu/clear/tdr/index.html

and an annotated bibliography is maintained on

http://www.geotdr.com

In addition there are several list servers that facilitate dialog among TDR specialists. ITI supports a general list server for all aspects of TDR use at

TDR-L@listserv.acns.nwu.edu

while there is another list server which addresses the needs of researchers involved principally in monitoring soil water content at

owner-sowacs@aqua.ccwr.ac.za

ORGANIZATION AND CONTENT OF THE BOOK

Chapters were kept small and focus on separate areas of application. Discussion of principles is separated from that of case histories of application for soil moisture and rock deformation because of the large amount of material and large number of examples of application. A short introductory paragraph and outline of topics for each chapter follows to more fully describe the contents. Rapidly changing information is included in Appendices such as Contact Information for Vendors, as well as Cable and Grout Properties.

CHAPTER 2. BASIC PHYSICS

TDR operates like radar in a coaxial cable. Distances to reflectors along the cable are calculated by time of flight, and characteristics of reflectors along or at the ends of a cable can be discerned from details of their respective reflected signals. This chapter describes the basic physics of signal generation, transmission, and attenuation along the coaxial cable. Also described are the physics of signals reflected by changes (both single and multiple) in the transmission path (along the cable) and by changes at special probes (at the end of the cable). Reflections produced along the length can lead to measurement of fluid levels and their presence as well as cable deformation produced anywhere along the cable. Those produced with specially constructed probes at the end of a cable can be employed to measure changes in moisture content of soils. Particular attention is paid to the physics of multiconductor probes and changes in dielectric constant because of their importance in the measurement of soil moisture or unsaturated water content.

CHAPTER 3. MONITORING SOIL MOISTURE

TDR sensitivity to changes in the dielectric constant of material between two conductors has been adapted to the measurement of moisture content. The success of this measurement technique with unsaturated soils has lead to its widespread adoption in agricultural study, irrigation control, and the study of unsaturated particulate material in general. The dielectric constant, $K = \epsilon/\epsilon_0$, of air is 1, ranges from 3 to 5 for most soil mineral grains, and is approximately 81 for water (at $20°C$). Thus, a small change in moisture content of unsaturated soils will have a significant effect on the bulk dielectric constant of the air-soil-water medium. The apparent dielectric constant is obtained by measuring the time for a voltage pulse to travel along the probe and return. This chapter describes probe design and procedures for their calibration as well as the variation in probe responses to changes in water content and soil mineralogy.

CHAPTER 4. FIELD EXPERIENCE AND VERIFICATION OF SOIL MOISTURE MEASUREMENT

The viability of TDR to measure water content of soils must be assessed by comparison with other techniques commonly employed. This chapter presents such comparative studies with lysimeters, Bowen ratio, neutron probes, and nuclear density gages. For the most part, these comparisons have been made with partially saturated soils.   After presentation of typical agricultural field validation studies, additional case studies are presented to demonstrate use of TDR to measure spatial and temporal changes in soil moisture associated with the performance of landfill covers, compacted fills, and pavement subgrades.

CHAPTER 5. MONITORING LOCALIZED DEFORMATION IN ROCK

Measuring rock deformation with TDR developed naturally from the original purpose of this technology, namely, identification of cable faults or deformities along telephone and power cables. TDR cables were first installed in holes drilled into rock masses around mining operations simply to locate cable breaks which correlated with the extent of complete rock mass failure. During one such application it was noticed that the TDR reflection waveforms changed incrementally with rock mass deformation before failure. Subsequent laboratory studies indicated that it was possible not only to quantify the magnitude of deformation but also, in some cases, to distinguish shearing from tensile deformation.   This chapter presents variations in waveform characteristics associated with cable deformation, cable calibration, and installation techniques for metallic cables installed in rock.

CHAPTER 6. FIELD EXPERIENCE AND VERIFICATION OF ROCK DEFORMATION MEASUREMENT

To date more than several dozen TDR deformation monitoring systems have been installed worldwide, which underscores its growing acceptance in the mining community. A number of case histories are presented to demonstrate the possibilities of quantitative evaluation of TDR waveforms for rock mass deformation. Comparisons are made with direct measurement of displacement in shallow boreholes, as well as those computed from inclinometer measurements. They are also made with indirect assessments based upon beam bending principles and cumulative horizontal displacements. This chapter closes with a number of case histories that illustrate the variety of environments wherein TDR measurement has been successfully employed.

CHAPTER 7. MONITORING SOIL DEFORMATION

It is possible to install a compliant TDR cable/grout system that will deform with localized soil shear zones in a manner similar to that observed with shearing of stiffer cable/grout systems by displacement along rock joints. In addition to developing the rationale for the use of compliant cable, this chapter presents several cases that demonstrate use of TDR cables in soil as well as weathered and soft rock. Remote monitoring of slope movement in soils has been inhibited by the manually operated nature of instruments commonly employed for this task, namely, the inclinometer or slope indicator. Inclinometers include down-hole electronics that must be manually lowered down a borehole and are thus difficult to operate remotely. TDR cable surveillance by its digital nature is automatically accomplished remotely and in addition is able to detect very thin shear zones.

CHAPTER 8. MONITORING STRUCTURAL DEFOR-MATION

This chapter presents the use of both metallic cable (MTDR) and optical fiber (OTDR) to monitor response of structures. While OTDR is not the focus of this book, a brief introduction to its physics and use is presented. Like its metallic counterpart, OTDR can be employed to monitor large volumes with a single fiber. While shorter wavelengths of optical signals allow measurement of elastic stress and strain, the location of the measurement must be known because the small strains can only be measured at selected locations. The chapter begins by describing use of MTDR for two aspects of structural response: 1) internal deformation, which involves monitoring localized response of individual structural members, and 2) external deformation, which involves monitoring the overall movement of a structure. It then closes with an overview of OTDR measurement of stress and strain within a structure.

CHAPTER 9. AIR–LIQUID INTERFACES

This chapter develops the background necessary for TDR measurement of fluid levels and discrimination of fluid types. TDR methods are compared with other approaches to measure pore water pressure near critical structures such as dams, as well as changes in ground water table elevation. Measurements are easily made via TDR because of the very large reflection that occurs at the air–water interface when water rises within the annular space of a hollow coaxial cable. In addition, TDR technology that identifies liquid presence is being extensively deployed to detect leakage of a variety of liquids over large areas or within large volumes. Leakage detection is possible with specially designed coaxial cables that selectively allow designated liquids to penetrate the outer conductor and permeate the dielectric material.

CHAPTER 10. ELECTRONICS

While various components of TDR systems can be acquired from a variety of suppliers, their functional relationship is constant and they can be described as a series of components. Since the state of the art in instrumentation is dynamic and is expected to be constantly changing, the attributes of each component are discussed in a generic manner.   Discussion in this chapter begins with the sensor/transducer components followed by connections from the sensors to the TDR pulser/sampler. Next, the system control methods are discussed, followed by components for storage and downloading of TDR data. Finally, power requirements for remote monitoring are addressed and detailed examples of systems that have been used to monitor rock deformation and soil moisture are summarized.

CHAPTER 11. SOFTWARE

The wide variety of software available for transmission and analysis of TDR signatures is summarized in this chapter. Some software is designed specifically for control of pulser/samplers and some specifically for analysis of TDR waveforms, while a third category serves both functions. In order to elaborate on the capabilities of available software, common details of programs are described. Discussion in this chapter begins with general control and acquisition software, followed by a description of programs that have been developed specifically for analysis of soil moisture measurements and cable deformation measurements.

Chapter 2
BASIC PHYSICS

Time Domain Reflectometry (TDR) is a remote sensing electrical measurement technique that has been used for many years to determine the spatial location and nature of various objects (Andrews, 1994). Radar is an early form of TDR, dating from the 1930s, with which most people are familiar. Radar consists of a radio transmitter which emits a short pulse of electromagnetic energy, a directional antenna, and a sensitive radio receiver. After the transmitter has radiated the pulse, the receiver then listens for an echo to return from a distant object, such as an airplane or ship. By measuring the time from the transmitted pulse until the echo returns and knowing the speed of light, the distance to the reflecting object may be easily calculated. Detailed analysis of the echo can reveal additional details of the reflecting object which aid in identification. The same principles that hold for radar also hold for metallic cable TDR and optical fiber TDR.

PULSE TESTING

There are a number of commercially available cable testers and the Tektronix 1500 series is a typical example. An ultra-fast rise time (200 psec) voltage step is launched into a test cable every 200 μsec as shown in Figure 2-1a (Tektronix, 1989a). When the pulses encounter a change in characteristic impedance (i.e., a cable fault), reflected pulses are returned to the cable tester.

Figure 2-1b is a block diagram of the TDR system. The TDR pulser generates a fast rise time step function. The step propagates through the sampling receiver and through the transmission line under test. The receiver uses an electronic sampling technique to produce a lower-frequency facsimile of the high frequency input. Many transmitted and reflected pulses are generated by the TDR in the time necessary to produce a scan. This scan is displayed as a reflection coefficient (i.e., ratio of reflected to

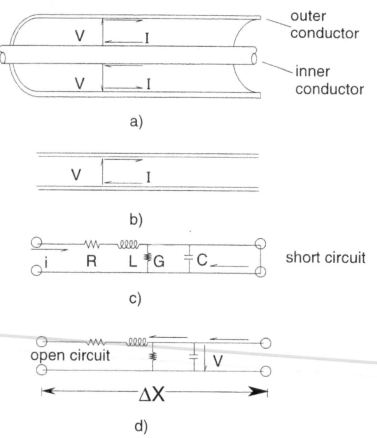

outer
conductor

inner
conductor

a)

b)

short circuit

c)

open circuit

d)

Figure 2-2.   Lumped transmission line parameters: (a) coaxial cable; (b) parallel wire transmission line; (c) lumped parameters with short circuit to define impedance; (d) lumped parameters with open circuit to define admittance.

inductance, L, in units of henries per meter;
3) the dissipation of energy by the two conductors is denoted by resistance, R, in units of ohms per meter;
4) the dielectric between the two conductors has a small conductivity, G, in units of siemens per meter which also dissipates energy.

The voltage V drops by an incremental amount, ΔV, in the incremental length, Δx. This voltage drop is caused by the current I flowing through the incremental resistance ($R * \Delta x$). Changes in the flowing current per unit time also decrease the voltage in proportion to the incremental inductance ($L * \Delta x$). That is,

$$-\Delta V = (R * \Delta x) * I + (L * \Delta x) * \partial I / \partial t \qquad (2\text{-}1)$$

Likewise, current, I, decreases by the increment, ΔI, due to the voltage difference across the incremental conductance ($G * \Delta x$). Changes in voltage per unit time also decrease the current in proportion to the incremental capacitance ($C * \Delta x$). That is,

$$-\Delta I = (G * \Delta x) * V + (C * \Delta x) * \partial V / \partial t \qquad (2\text{-}2)$$

Using Laplace transformation techniques, Equations 2-1 and 2-2 can be solved for equations in the form of current only and voltage only (Dworak et al., 1977),

$$\frac{\partial^2 I}{\partial^2 x} - (Ls + R)(Cs \cdot G) \, I = 0 \qquad (2\text{-}3)$$

$$\frac{\partial^2 V}{\partial^2 x} - (Ls + R)(Cs + G) \, V = 0 \qquad (2\text{-}4)$$

where s is the Laplace transform operator.

Spergel (1972) summarizes a simpler approach using the concept of impedances with equivalent circuits to arrive at the transmission line parameters that appear in Equations 2-3 and 2-4. It is desirable to analyze the series and parallel elements of the equivalent circuit separately. Kirchhoff's law allows us to add impedances in series and admittances in parallel configuration. The impedance, Z, of the equivalent circuit can be measured as shown in Figure 2-2c with the output short-circuited. The parallel components are shorted; only the series components are measured. Impedance of the inductor is a function of time so it can be out of phase with the resistive impedance. Since the sum of voltages must be zero,

$$ZI=RI+j\omega LI \tag{2-5a}$$

where R and L are lumped values for the length, Δx, and complex notation is a consequence of the phase relationship. Impedance can then be expressed as

$$Z=R+j\omega L. \tag{2-5b}$$

Information about admittance, Y, can be gained by measuring from the other end when the front end is open-circuited. Figure 2-2d shows the admittance measurement setup. Since the series elements are left open, only the parallel components will be measured. Admittance of the capacitor is a function of time so it can be out of phase with the admittance of the dielectric conductance. Since the sum of currents at a junction must be zero,

$$YV=GV+j\omega CV \tag{2-6a}$$

where G and C are lumped values for the length, Δx, and complex notation is again a consequence of the phase relationship. Admittance is then expressed as

$$Y=G+j\omega C. \tag{2-6b}$$

Two important parameters can be derived from these impedance and admittance expressions.

The transmission line parameters in Equations 2-3 and 2-4 can be written as

$$\gamma = \sqrt{ZY} = \sqrt{(R + j\omega L)(G + j\omega C)} \qquad (2\text{-}7)$$

which is called the propagation constant of the transmission line. By definition, another parameter is the characteristic impedance of the transmission line in units of ohms,

$$Z_o = \sqrt{\frac{Z}{Y}} = \sqrt{\frac{R + j\omega L}{G + j\omega C}} \; . \qquad (2\text{-}8)$$

The circular frequency $\omega = 2\pi f$ so that for radio frequencies ($f = 0.3$ MHZ to 30 GHz), $R \ll \omega L$ and $G \ll \omega C$. So, the transmission line parameters are dominated by L and C, and the propagation constant in Equation 2-7 becomes

$$\gamma = \sqrt{LC}. \qquad (2\text{-}9)$$

Its inverse is the propagation velocity with which an electromagnetic wave can travel along the transmission line

$$V_p = \frac{1}{\sqrt{LC}} \qquad (2\text{-}10)$$

which is expressed as a percentage of the speed of light (3×10^8 m/sec). The characteristic impedance of a transmission line from Equation 2-8 is

$$Z_o = \sqrt{\frac{L}{C}} \, . \tag{2-11}$$

TDR REFLECTION TYPE

The type of waveform observed with a TDR cable tester depends on several factors, including the type of cable fault. It is a major advantage of TDR that you can locate faults by virtue of the travel time and also identify the type of cable fault. Faults are basically changes in the basic transmission line properties (R, C, and L) which can be measured as changes in impedance (Z) using TDR. For this discussion we will assume that the generator produces a step function pulse as shown in Figure 2-1a.

RESISTIVE TERMINATIONS AND THE REFLECTION COEFFICIENT

TDR waveforms observed for various pure resistive terminations, $Z_t = R_t$, are shown in Figure 2-3a (Andrews, 1994). If a 50 ohm termination is connected to the end of a 50 ohm coaxial cable, $Z_t = Z_o$, no reflection occurs and the TDR display on the scope is a flat line. If Z_t is greater than Z_o, then a positive step is observed. For $Z_t < Z_o$, a negative step is observed. The actual value of Z_t may be calculated from the size of the incident step, V_i, and the reflected pulse, V_r. The reflection coefficient, ρ or rho, is defined as

$$\rho = V_r / V_i \, . \tag{2-12}$$

Note that V_r may have either a positive or negative value and likewise for ρ.

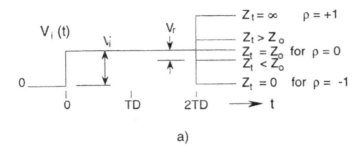

$$Z_t = \infty \qquad \rho = +1$$
$$Z_t > Z_o$$
$$Z_t = Z_o \quad \text{for} \quad \rho = 0$$
$$Z_t < Z_o$$
$$Z_t = 0 \quad \text{for} \quad \rho = -1$$

a)

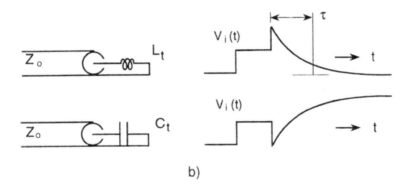

b)

Figure 2-3. TDR waveforms for ideal resistive and reactive terminations: (a) resistive; (b) reactive (from Andrews, 1994).

When Equation 2-4 is solved for voltage, V, as a function of distance, the solution contains the term (Dworak et al., 1977)

$$\rho = (Z_t - Z_o)/(Z_t + Z_o) \tag{2-13}$$

which is the reflectivity due to a load impedance mismatch. This is equivalent to the reflection coefficient in Equation 2-12. Rearranging terms,

$$Z_t = Z_o * (1 + \rho)/(1 - \rho). \tag{2-14}$$

For the matched case, $Z_t = Z_o$ and $\rho = 0$. For an open circuit, $Z_t = \infty$ ohms and $\rho = +1$. For a short circuit, $Z_t = 0$ ohms and $\rho = -1$. Equations 2-12 through 2-14 hold for both pure resistive terminations and connections to other transmission lines of different characteristic impedances. These reflective properties are also found for one-dimensional compressive stress wave propagation along rods and beams.

REACTIVE TERMINATIONS

Reactive components can also be measured using TDR. Figure 2-3b shows the TDR waveforms obtained for terminations of simple inductors or capacitors (Andrews, 1994). The inductor initially appears as an open circuit to the fast-rising edge of the TDR step pulse (i.e., the high frequencies). Thus it initially gives a $\rho = +1$. Later in time, the inductor appears as a short circuit to the flat top of the TDR step pulse (i.e., the dc portion). Thus the final TDR value is $\rho = -1$. The capacitor performs exactly the opposite with an initial $\rho = -1$ and final $\rho = +1$. The L_t and C_t value may be determined by measuring an exponential time constant of the TDR response (Halliday and Resnick, 1962)

$$L_t = Z_o * \tau_L \tag{2-15}$$

$$C_t = \tau_c / Z_o \tag{2-16}$$

where τ_L and τ_c are the inductive and capacitive time constants, respectively. The exponential time constant is the time required for ρ to decrease from its peak value to 63% $(1 - e^{-t/\tau} = 1 - e^{-1} = 0.63)$ of this value.

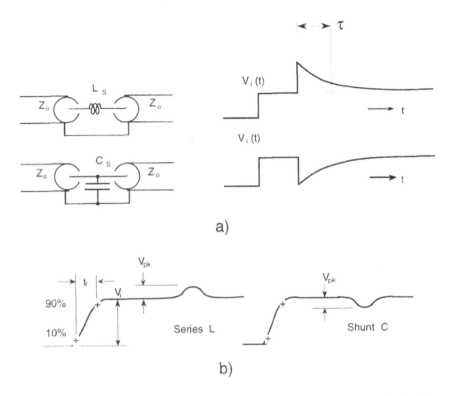

**Figure 2-4. TDR waveforms for reactances along coaxial cable: a) ideal series
L and shunt C; b) smaller L and C reactances (from Andrews, 1994).**

REACTANCES ALONG CABLE

Figure 2-4a shows a very common situation encountered when
dealing with coaxial cables. These are the cases of either a series
inductance, L_s, or a shunt capacitance, C_s, within the coaxial cable.
This could be caused by a poor connector, for example. Again by
measuring the appropriate time constant, τ_L or τ_c, the value of L_s
or C_s may be determined (Andrews, 1994)

$$L_s = (2*Z_o)*\tau_L \qquad\qquad (2\text{-}17)$$

$$C_s = \tau_c / (Z_0/2). \tag{2-18}$$

RISE TIME

The TDR waveforms shown in Figure 2-3b were for the case when the TDR waveform is ideal with a rise time of zero seconds. As a practical matter, pulse generators and samplers have a finite rise time and these waveforms do not have sharp corners but have smooth rounded corners. For large inductors or capacitors, their time constants will be much greater than the system rise time, t_r, and the leading edge will still produce a reflection coefficient of +1 or −1, respectively. No visible reflections will occur for tiny inductors or capacitors with time constants much less than the TDR system rise time, t_r. Thus they cannot be measured (Andrews, 1994; Hewlett-Packard, 1988).

Figure 2-4b shows the TDR displays that will occur for inductors and capacitors when the time constant associated with the reactive components is the same order of magnitude as the TDR system risetime ($\tau \approx t_r$). For these smaller reflections the maximum or minimum reflection coefficient is always less than +1 or −1. These smaller inductors or capacitors can be detected as shown by V_{pk} in Figure 2-4b, even though their responses are not the same as shown in Figure 2-3b.

For reflections in which V_{pk} is less than 10% of V_i, the reactive components L_s and C_s can be approximated as (Cole, 1976)

$$L_s = (2*Z_o)*t_r*[V_{pk}/(0.8*V_i)] \tag{2-19}$$

$$C_s = (2/Z_o)*t_r*[V_{pk}/(0.8*V_i)] \tag{2-20}$$

ELECTRIC FIELD IN COAXIAL CABLE

As an electromagnetic pulse is transmitted along a coaxial cable, it is constantly being transformed from an electric field to a magnetic field and vice versa. Consider a transverse cross section of the cable at one instant in time. The electric field associated with a electromagnetic pulse as it passes through the plane of this cross section is equivalent, mathematically, to a laminar flow field produced by steady state seepage. Both fields consist of perpendicular equipotential and flow (or stream) lines as shown by the comparison of water flowing below a dam and current flowing between the inner and outer conductor of a coaxial cable in Figure 2-5. Thus the mathematics of a potential field can be employed to link experience in these two diverse fields using the nomenclature in Table 2.1.

   The flow nets shown in Figure 2-5 are graphical representations of the potential field at some instant in time. The rules for drawing a potential field flow net can be translated between the languages of electrostatics and steady-state seepage as follows:

   1) An electrostatic potential difference (analogous to a pressure difference) exists between the concentric conductors.

   2) Dissipation of the electrostatic potential difference depends on the dielectric constant and specific conductivity of material between the conductors (analogous to hydraulic permeability).

   3) The electric field lines (which are analogous to fluid flow lines) begin and terminate on potential surfaces.

   4) The streamlines are necessarily at right angles to the equipotential lines. The electric force (analogous to fluid velocity) at every point is proportional to the space gradient of the electrostatic potential. The space gradient is zero along a potential line.

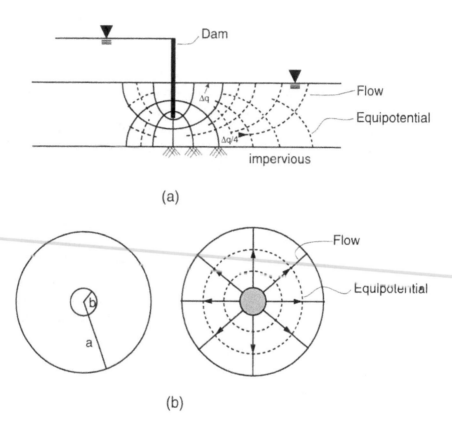

(a)

(b)

Figure 2-5.   Analogy between flow through porous media and electromagnetic wave propagation in coaxial cable: (a) flow net for water seeping beneath a dam; (b) flow net for current flow from inner to outer connector (from Su, 1987).

Table 2.1. Correspondences between the flow of an incompressible liquid through a porous medium, electrostatics, and current conduction (after Cedergren, 1967; reprinted by permission of John Wiley & Sons, Inc.).

Hydrodynamics of steady-state flow through porous media (incompressible liquids)	Electrostatics	Current conduction
Pressure: p	Electrostatic potential: ϕ	Voltage (potential): V
Negative pressure gradient: $-\nabla p$	Electric field strength vector: $\underline{E} = -\nabla\phi$	Negative potential gradient: $-\nabla V$
Permeability/Viscosity: k/μ	Dielectric constant/4π: $\epsilon/4\pi$	Specific conductivity: σ
Velocity vector: $\underline{v} = -(k/\mu)\nabla p$	Dielectric displacement: $\epsilon/4\pi\ \underline{E} = -\epsilon/4\pi\ \nabla\phi$	Current vector: $\underline{I} = -\sigma\nabla V$
(Darcy's law)	(Maxwell's law of dielectric displacement)	(Ohm's law)
Equipressure surface: p = constant	Equipotential surface: ϕ = constant	Equipotential surface: V = constant
Impermeable boundary or streamline: $\partial p/\partial n = 0$	A tube or line of force: $\partial\phi/\partial n = 0$	Free or insulated surface of tube or line of flow: $\partial V/\partial n = 0$

CABLE GEOMETRY AND CAPACITANCE

For the coaxial cable shown in Figure 2-5b, the capacitance per unit length, C, and inductance per unit length, L, are functions of cable geometry (Halliday and Resnick, 1962),

$$C = 2\pi\epsilon/\ln(a/b) \qquad (2\text{-}21)$$

$$L = (\mu/2\pi)\ln(a/b) \qquad (2\text{-}22)$$

where a and b are the radii of the outer and inner conductors; μ and ϵ are the magnetic permeability and dielectric permittivity, respectively, of the material between the conductors. Therefore, the impedance, Z, is also a function of cable geometry. Substituting Equations 2-21 and 2-22 into 2-11,

$$Z = \sqrt{\frac{L}{C}} = \frac{1}{2\pi}\ln(\frac{a}{b})\sqrt{\frac{\mu}{\epsilon}}. \qquad (2\text{-}23)$$

This geometric dependence can be summarized as

$$Z = f(a/b) \qquad (2\text{-}24)$$

$$C = f(a/b). \qquad (2\text{-}25)$$

Using finite element techniques, Su (1987) approximated the effect of cable deformation on capacitance by computing an equivalent capacitance per unit length for different deformed geometries.   The deformed geometries were observed during laboratory shear tests on cables. The outer conductor of the coaxial cable was deformed sharply while the inner conductor, separated from the outer conductor by a relatively soft dielectric material, was deformed more gradually over a longer length of the cable. The capacitance for a coaxial cable was estimated using flow net techniques developed for estimating total seepage flow such as shown in Figure 2-5 (Cedergren, 1967),

$$C = f(N_Q/N_v, \epsilon, \mu) \qquad\qquad (2\text{-}26)$$

where N_Q is the number of streamlines and N_v is the number of equipotential lines. If the dielectric permittivity, ϵ, is constant, then C is proportional to N_Q/N_v, which is referred to as the "shape factor" for the particular flow net. An equivalent lumped capacitance for the deformed transverse cross section was estimated by multiplying the computed capacitance per unit length, C, times a deformed length equal to twice the cable diameter based on the observed deformed cables.

Su (1987) designated the impedance and capacitance of the original cable cross section as Z_o and C_o, respectively, and assumed that μ and ϵ did not change as the cable was deformed. He designated impedance and capacitance of the deformed cross section as Z_1 and C_1 and, based on Equations 2-24 and 2-25, proposed that the ratio Z_1/Z_0 is linearly proportional to the ratio C_0/C_1,

$$Z_1 = (C_o/C_1)Z_o. \qquad\qquad (2\text{-}27)$$

Figure 2-6 compares calculated and measured capacitance change (as a percentage of C_o) versus shear deformation for a 12.7 mm diameter coaxial cable. As shown, the increase in equivalent capacitance is proportional to the increase in deformation of the transverse cross section. This increase in capacitance produced an increase in the TDR reflection coefficient, ρ.

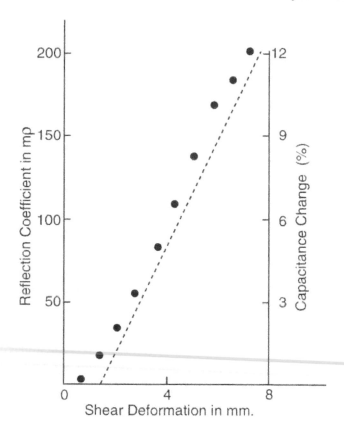

Figure 2-6. Capacitance change for 1.5 m long cable subjected to shear deformation; solid dots from numerical model; dashed line from trend of laboratory measurements (from Su, 1987).

MEASUREMENT OF DIELECTRIC WITH PARALLEL RODS

The TDR technique of measuring electrical properties of materials was introduced by Fellner-Feldegg (1969) using alcohols in coaxial cylinders. Topp et al. (1980) extended this application to earth materials by using TDR to determine the volumetric water content of soils in coaxial sample holders. The coaxial cylinders were not

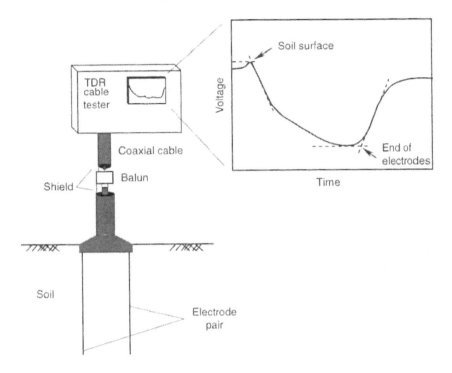

Figure 2-7. System for measurement of dielectric constant of material between two parallel rods (from Herkelrath et al., 1991).

suitable for field measurements, so Topp and Davis (1985) used a transmission line consisting of parallel rods based on the work of Davis and Chudobiak (1975). The basic TDR circuit is illustrated in Figure 2-7. The TDR trace, as displayed on the oscilloscope screen, develops as follows: the cable tester applies a fast rise-time voltage step to the 50-ohm coaxial cable and triggers a sampler. The step pulse travels down the coaxial cable, through an impedance-matching transformer (or balun), and into a shielded two-wire television cable (a balanced line). The signal travels along the cable until it reaches the soil electrodes, where, because there is an impedance mismatch at the soil surface, part of the signal is reflected back toward the cable tester and part of the signal passes along the parallel rods in the ground. The signal traveling in the ground ultimately reaches the end of the rods, sending a second reflection back toward the cable tester. By repeating this process many times, a stable waveform is developed as shown in the insert of Figure 2-7. This waveform reveals the time between arrival of reflections from the soil surface and from the end of the rods, and the magnitudes of these reflections.

PROPAGATION VELOCITY AND DIELECTRIC CONSTANT

A discussion of TDR applied to soils can be found in Davis and Chudobiak (1975). Briefly, the propagation velocity, V_p, of a voltage pulse traveling along a transmission probe of a known length, l_p, is

$$V_p = 2 l_p / t \qquad (2-28)$$

where t is the time between reflections from the top and bottom of the probe, and $2 l_p$ indicates that the pulse travels twice (i.e., down and back) along the probe.

In order to convert this measured propagation velocity to a material property, Topp et al. (1980) considered electrical properties of heterogeneous materials. The velocity at which an electromagnetic wave can travel through a medium is dependent on the permittivity ($\epsilon = K \epsilon_0$) of the medium

$$V_p = c / \epsilon^{1/2} \qquad (2-29)$$

where c is the speed of an electromagnetic wave in free space ($3 * 10^8$ m/sec). This is equivalent to the definition of propagation velocity given in Equation 2-10, but it is derived from electromagnetic wave theory rather than transmission line theory.

Rearranging Equation 2-29 to define dielectric permittivity as a function of the propagation constant,

$$\epsilon = (c/V_p)^2. \tag{2-30}$$

The dielectric permittivity is actually a complex quantity, as discussed in Chapter 3,

$$\epsilon = \epsilon' - j\epsilon'' \tag{2-31}$$

where the real part, ϵ', is a measure of the polarizability of the material constituents (Hilhorst and Dirksen, 1994). The imaginary part, ϵ'', represents energy absorption by dielectric losses and ionic conduction. Even though the effects of dielectric loss were not measured, Topp et al. (1980) felt that these did exist in their estimate of ϵ' and called the measured dielectric "constant" an apparent dielectric constant

$$K_a = \epsilon' \tag{2-32}$$

which they substituted for ϵ in Equation 2-30. Note that $\epsilon_o = 1$ in the CGS system. Inserting V_p from Equation 2-28, the apparent dielectric constant measured with TDR is

$$K_a \approx (c/V_p)^2 = [(ct)/(2l_p)]^2. \tag{2-33}$$

ELECTRIC FIELD DISTRIBUTION AND SAMPLING VOLUME AROUND PARALLEL RODS

A coaxial cable transmission line essentially constrains the electromagnetic field within the annulus between the inner and outer conductors (Figure 2-5). Such a constraint does not exist in the case of a parallel rod transmission line (Figure 2-7). As the size of a parallel rod probe (length and spacing) is increased, the resolution of the TDR system decreases. Thus compromises must be made depending on the specific measurement required. Topp and Davis (1985) found that parallel-rod transmission lines with center-to-center spacing of about 50 mm are a practical compromise. Their work showed that the volume of soil measured by TDR was essentially a cylinder whose axis lies midway between the rods and whose diameter is 1.4 times the spacing between the rods. This gave a cross section of 3800 mm^2 for rods spaced 50 mm apart.

Baker and Lascano (1989) investigated the spatial sensitivity of parallel rod probes using a laboratory procedure in which the distribution of water surrounding the waveguides was controlled to give different spatial arrangements of water around a probe. Stainless steel rods 3.2 mm in diameter and 300 mm long were spaced 50 mm apart in a matrix of glass tubes. The sensitivity was found to be largely confined to a region with a cross-sectional area of approximately 1000 mm^2 (20 mm by 65 mm) surrounding the waveguides, although a limited sensitivity extends much farther, enclosing 3600 mm^2 (60 mm by 80 mm) which was consistent with Topp and Davis (1985). The width of the region of sensitivity normal to the plane containing the parallel rods is approximately 30 mm (1.7 times spacing). Finally, there was no discernible variation in sensitivity longitudinally, i.e., along the length of the waveguide. The data indicated that sensitivity ends abruptly at the ends of the waveguides, i.e., changes in water content just beyond the end of the waveguides have no discernible effect on the signal.

Zegelin et al. (1989) investigated the field distributions of three- and four-rod probes which were developed to provide increasingly better approximations to an ideal coaxial cell. The potential distributions in Figure 2-8 do not tell us precisely what volumes of soil are sampled by the probes nor how information from various parts of the dielectric medium penetrated by the field is weighted in the resultant signal. They do, however, substantiate the claim that the probes emulate coaxial cells as seen by comparing the 4-wire field in Figure 2-8 with the equipotential field in Figure 2-5.

Knight et al. (1994) used the theory of propagation of electromagnetic waves along transmission lines to investigate the distribution of electromagnetic energy around parallel rods. They defined a relative spatial sensitivity function for the energy distribution around a central rod. As the number of rods becomes large,

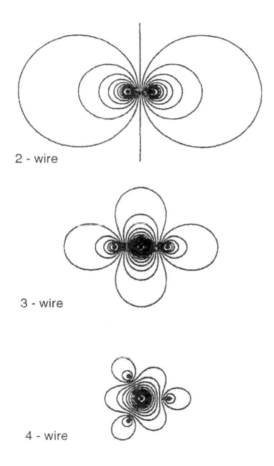

2 - wire

3 - wire

4 - wire

Figure 2-8. Contours of dimensionless electric field distribution normal to direction of probe insertion for a material of uniform dielectric constant (from Zegelin, White, and Jenkins, 1989).

the average relative density approaches the distribution of a coaxial cell, and most of the energy is inside a unit radius. Figure 2-9a shows contours in the first quadrant of the x-y plane of the relative spatial sensitivity function for a 3-rod probe with one central rod and two outer rods. The out-of-plane (y-direction) spread of energy into the far field is less for a 3-rod probe than for a 2-rod probe (0.7 times the rod spacing vs. 1.7 times the spacing). To indicate the effect of increasing the number of rods, Figure 2-9b shows contours of the spatial sensitivity function for a nine-rod probe. Shielding by the eight outer rods caused most of the energy to be contained in the region defined by a unit radius.

MEASUREMENT OF CONDUCTIVITY WITH PARALLEL RODS

For a conducting medium such as a saline soil, the launched pulse is attenuated due to conduction losses into the medium (Dalton and van Genuchten, 1986). The amplitude of the reflected voltage pulse is thus diminished in proportion to the electrical conductivity. Figure 2-10 shows parameters of a TDR waveform obtained with a parallel two-rod probe. Dalton et al. (1984) conducted a study to determine if attenuation of the launched signal in a conductive medium could be used to measure soil salinity. If the measurements of transit time, t, and attenuation are sufficiently independent, then the conductivity and dielectric constant can be determined simultaneously. In Figure 2-10, V_o represents the output of the pulse generator, V_1 is the magnitude of the voltage pulse that enters the parallel-rod waveguide, and V_2 designates the magnitude of the reflected wave when it has returned to the waveguide input (Fellner-Feldegg, 1969). Note that there is an exponential decrease from V_o to V_1. Zegelin et al. (1989) show that with improved probe designs, this decrease occurs rapidly and remains constant over the probe length.

V_2 must be greater than zero in order to be able to distinguish the time interval, t. In a nonconducting medium, V_2 will be equal to V_1 and t is easily measured. However, when the medium

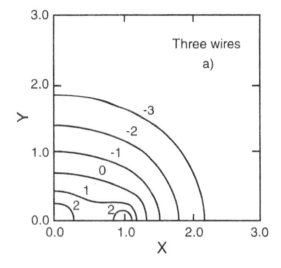

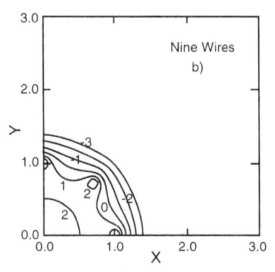

Figure 2-9. Contours of the relative spatial sensitivity function: (a) three prong probe; (b) nine prong probe (from   Knight, White,   and Zegelin, 1994).

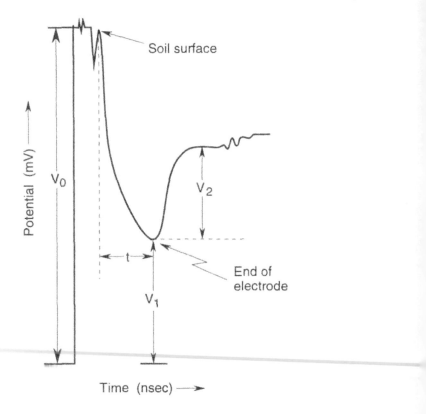

Figure 2-10. Parameters of a TDR waveform: output voltage, V_o; voltage transmitted along probe, V_1; time for pulse to travel along probe and return, t; and reflected voltage, V_2 (from Dalton et al., 1984).

is conductive, as in a saline soil, the amplitude of V_2 will be attenuated in proportion to the conductivity. The attenuation of V_2 can be used to measure the electrical conductivity of the medium and, therefore, the salt content.

ATTENUATION IN CONDUCTIVE MEDIA

The interaction and propagation of electromagnetic waves in conductive and dispersive media are extremely complex physical phenomena (Dalton, 1992). An empirically based design of transmission lines was in existence in the mid-1800s and since then electrical engineers have developed a very sophisticated method of analysis of wave propagation based on a combination of lumped circuit analysis, distributed circuit analysis, and direct solution of Maxwell's equations with appropriate boundary conditions. These concepts form the basis of most analyses applied to waveguide electromagnetic propagation (Ramo et al., 1965). The direct application of standard transmission line theory to earth materials is not usually adequate, because most applications of current theory are applied to the study of materials that have low electrical conductivities. Dalton et al. (1984) developed an analysis technique using a mixture of approaches.

According to electromagnetic field theory, the amplitude of a signal traveling a distance $2l_p$ in an electrically conductive medium with an attenuation coefficient, α, will be diminished exponentially and, assuming perfect reflection, the reflected voltage returning to the source is given by

$$V_2 = V_1 \exp(-\alpha 2 l_p). \qquad (2\text{-}34)$$

An approximate relation for α applicable to steady-state, sinusoidal input conditions and low-loss lines is given by (Ramo and Whinnery, 1959; Dalton and van Genuchten, 1986)

$$\alpha = \frac{R\sqrt{C}}{2\sqrt{L}} + \frac{G\sqrt{L}}{2\sqrt{C}}. \qquad (2\text{-}35)$$

Note that R and G were neglected for purposes of defining characteristic impedance of a transmission line in Equation 2-11. However, Equation 2-35 incorporates these parameters since TDR measurement of conductivity utilizes attenuation of the low-frequency and dc components of the transmitted and reflected voltage pulses. Expressions of L, C, and G for parallel rod transmission lines which are more appropriate for a discussion of attenuation are (*Am. Inst. of Phys. Handbook*, 1957)

$$L = \frac{\mu'}{\pi} \cosh^{-1}(s/d) \tag{2-36}$$

$$C = \frac{\pi \epsilon'}{\cosh^{-1}(s/d)} \tag{2-37}$$

and

$$G = \frac{\pi \sigma}{\cosh^{-1}(s/d)} \tag{2-38}$$

in which σ is the electrical conductivity of the medium (siemens/meter); μ' is the magnetic permeability of the medium, $\mu\mu_o$, where μ_o is the magnetic permeability of free space ($\mu_o = 4\pi * 10^{-7}$ henries/meter) and μ is the relative magnetic permeability of the medium; ϵ' is the dielectric permittivity of the medium, $\epsilon\epsilon_o$, where ϵ_o is the dielectric permittivity of free space ($\epsilon_o = 10^{-9}/36\pi$ farads/meter) and ϵ is the relative dielectric permittivity of the medium, while s and d are the electrode separation and diameter (meter). Substituting Equations 2-36 through 2-38 into Equation 2-35

$$\alpha = \sigma/2[(\mu\mu_o)/(\epsilon\epsilon_o)]^{1/2}. \tag{2-39}$$

For soils low in magnetic material, $\mu=1$, and substituting values of ϵ_o and μ_o, Equation 2-39 becomes

$$\alpha = 60\pi\,\sigma/\epsilon^{1/2}. \qquad (2\text{-}40)$$

Combining Equations 2-40 and 2-34 yields a technique for TDR measurement of the medium's electrical conductivity

$$\sigma = [\epsilon^{1/2}/(120\pi l_p)]\ \ln(V_1/V_2). \qquad (2\text{-}41)$$

Under conditions for which travel time, t, is measurable as shown in Figure 2-10, ϵ ($\approx K_a$) is measured using Equation 2-33. Equation 2-41 can then be used to calculate σ from the TDR waveform.

In order to compare the TDR determination of σ values as predicted from Equation 2-41 with those measured using a conductivity meter, Dalton et al. (1984) brought ten soil columns to equal water contents, using waters of ten different σ values. Twenty-five days after infiltrating each column with waters of known electrical conductivities, they made TDR measurements and typical waveforms are shown in Figure 2-11. Consistent with Equation 2-34, voltage returned to the cable tester, V_2, was attenuated as the conductivity increased.

ATTENUATION OF MULTIPLE REFLECTIONS

Topp et al. (1988) and Yanuka et al. (1988) have pointed out that V_1 and V_2 as shown in Figure 2-10 are values at the cable tester interface and that the effects of the intervening coaxial cable and impedance matching transformer are neglected (Dalton, 1992). Yanuka et al. (1988) with the aid of a ray diagram give a detailed account of the transmission and reflection of an electromagnetic pulse impinging on a layered medium as shown in Figure 2-12. Based on this analysis, Topp et al. (1988) calculate the voltage after

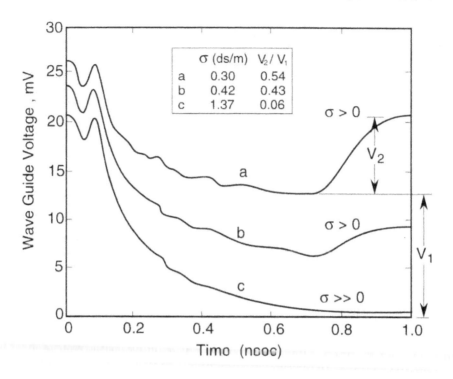

Figure 2-11.   TDR signal attenuation in conductive media showing effect of increasing conductivity (from Dalton et al., 1984 and Dalton and van Genuchten, 1986).

one return trip as

$$V_1 = V_o(1 + \rho) \tag{2-42}$$

and after two roundtrips,

$$V_2 = V_o(1 + \rho) + V_o(1 - \rho^2)e^{-\alpha l_p}, \tag{2-43}$$

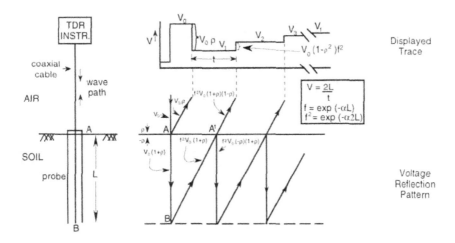

Figure 2-12. Schematic representation of multiple TDR reflections in soil showing: TDR system (on left), idealized trace (top right), and voltage reflection pattern (bottom right) (from Topp et al., 1994).

where ρ is the reflection coefficient, V_o is the voltage output from the TDR, and α is the attenuation constant for which Topp et al. (1988) adopted the definition given in Equation 2-40. Estimates of V_o, V_2, and ϵ ($\approx K_a$) from a TDR trace such as given in Figure 2-12 allow calculation of α which may then be used to obtain σ_{eff} (Zegelin et al., 1989),

$$\sigma_{eff} = \frac{\sqrt{K_a}}{120\pi l_p} \ln\left[\frac{V_1(2V_o - V_1)}{V_o(V_2 - V_1)}\right] \tag{2-44}$$

where K_a is obtained by Equation 2-33.

Topp et al. (1988) pointed out that the inclusion of V_2 poses a problem experimentally since frequency-dependent attenuation makes it a somewhat arbitrary measure. To overcome this, Yanuka et al. (1988) introduced a multiple reflection model and used the amplitude of the signal, V_f, after all reflections within a coaxial sampler holder had taken place. The model equations can be rearranged as (Zegelin et al., 1989)

$$\sigma_{eff} = \frac{\sqrt{K_a}}{120\pi l_p} \ln\left[\frac{V_1 V_f - V_o(V_1 + V_f)}{V_o(V_1 - V_f)}\right]. \qquad (2\text{-}45)$$

VARIETY OF TDR WAVEFORMS

The potential applications of TDR can lead to a variety of measurements in one monitoring program. Thus, the system may be required to record a wide variety of signal types. For example, assume that a project requires monitoring water level, rock/soil shearing, and soil moisture simultaneously. A multiplexing system can be deployed to fulfill this requirement with a single TDR pulser. A 100 m long low loss cable could be attached to the first channel. At the end of the low loss lead cable, a 10 to 20 m long air dielectric coaxial cable could be employed as a probe to measure water level changes. A 300 m long coaxial cable could be connected to the second channel to measure rock/soil shearing. Similarly, coaxial cable could be attached to a third channel to monitor soil moisture. Figures 2-10, 2-11, and 2-13 show the variety of waveforms that would be acquired in this situation. These figures illustrate waveforms from measurement of soil moisture (2-10), change in conductivity (2-11), water within a coaxial cable (2-13a), and localized shearing of rock or soil (2-13b). Each channel would be programmed for different acquisition settings to be compatible with the variety of waveforms.

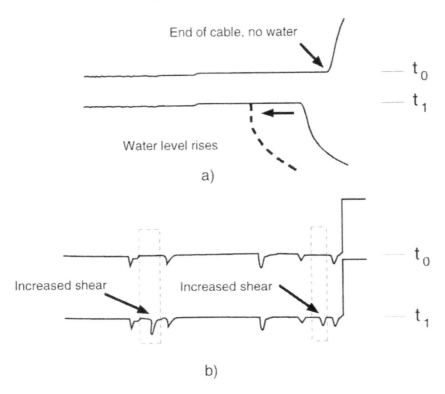

End of cable, no water

Water level rises

a)

Increased shear   Increased shear

b)

Figure 2-13. Change in TDR waveforms with change in measured parameter: (a) hollow coaxial cable in water; (b) coaxial cable sheared by rock or solid mass. See Figures 2-10 and 2-11 for waveforms obtained with probes such as the one in Figure 2-7 that is used to measure changes in soil moisture and conductivity.

Chapter 3
MONITORING
SOIL MOISTURE

TDR sensitivity to changes in the dielectric constant of material between two conductors has been adapted to the measurement of moisture content. The success of this measurement technique with unsaturated soils has lead to its widespread adoption in agricultural study, irrigation control, and the study of unsaturated particulate materials in general. The dielectric constant, $K=\epsilon/\epsilon_o$, of air is 1, ranges from 3 to 5 for most soil mineral grains, and is approximately 81 for water (at 20°C). Thus, a small change in moisture content of unsaturated soils will have a significant effect on the bulk dielectric constant of the air–soil–water medium. A typical system consists of a standard TDR cable tester with a serial interface, a multiplexer for use with multiple probes, a computer, and software for data storage and analysis. Probes should be calibrated in air, water, and the soil prior to field installation to correct for characteristics of the material between the probe conductors and any unusual soil mineralogy. The apparent dielectric constant is obtained by measuring the time for a voltage pulse to travel along the probe and return. This chapter describes probe design and procedures for their calibration as well as the variation in probe responses to changes in water content and soil mineralogy.

SOIL MOISTURE TERMINOLOGY

There are three common methods used to express soil moisture content. Volumetric water content, θ_v, can be measured indirectly by TDR travel time, while the two gravimetric water contents, θ_m and w, must be measured by oven drying samples in the laboratory,

$$\theta_v = V_w/V_{tot} \tag{3-1}$$

where V_w is the volume of water in the total volume V_{tot},

$$\theta_m = W_w/W_{tot} \tag{3-2}$$

where W_w is the weight of water and W_{tot} is the total weight of the moist soil, and

$$w = W_w/W_s \tag{3-3}$$

where W_s is the weight of dry solids. Gravimetric water content, θ_m, can be converted to volumetric water content using the bulk density of the moist soil, ρ_b, and the density of water, ρ_w (= 1 gm/cm^3 = 1 Mg/m^3),

$$\begin{aligned}\theta_m &= W_w/W_{tot} = \gamma_w V_w/(\gamma_{bulk} V_{tot}) = (\gamma_w/\gamma_{bulk})\theta_v \\ &= (\rho_w g/\rho_b g)\theta_v = (\rho_w/\rho_b)\theta_v \end{aligned} \tag{3-4}$$

The gravimetric water content, w, is used most commonly,

$$\begin{aligned}w &= W_w/W_s = \gamma_w V_w/(\gamma_d V_{tot}) = (\gamma_w/\gamma_d)\theta_v \\ &= (\rho_w/\rho_d)\theta_v \end{aligned} \tag{3-5}$$

where the dry density, $\rho_d = \gamma_d/g$ and w is related to the gravimetric water content, θ_m,

$$w = (\rho_w/\rho_d)\theta_v = (\rho_w/\rho_d)(\rho_b/\rho_w)\theta_m$$
$$= (\rho_b/\rho_d)\theta_m \ .$$

(3-6)

PROBE DESIGN

The three-prong probe shown in Figure 3-1 is just one of several configurations which have been used by researchers and practitioners. Among the factors which have been considered during the continuing evolution of probe design are probe length, rod diameter, rod spacing, sampling volume, connector type, connector corrosion, and probe robustness. Larger probes were developed for use in pavement subgrades where they must sustain harsh treatment during subgrade placement and compaction. Another design developed by Yong and Hoppe (1989) consisted of a 3-mm-diameter, 20-mm-long brass rod extending from the center conductor of a coaxial cable. Janoo et al. (1994) reported problems with corrosion of the connections between cables and probes when they were buried and this was corrected by encapsulating the connections in epoxy.

If the probes are too short, then the length of the TDR reflections at the beginning and end of the probe is long relative to the total waveform length (refer to Figure 2-7 and Figure 10-2) and the travel time used in Equation 2-33 is determined less accurately. Longer probes increase the travel time, but attenuation also increases which makes it difficult to detect the end-of-probe reflection. Attenuation is also a problem if lead cables used with the probes are too long.

An important question when using this technique is: What is the region of influence around TDR rods? As discussed in Chapter 2, comparison of the electric field distribution around a three-rod probe and that around a two-rod probe shows that the three-rod probe is almost twice as sensitive since the area which has the greatest influence on the measurement is associated with regions where the potential gradient is greatest (Whalley, 1993). Researchers have considered two approaches to assessment of the region of

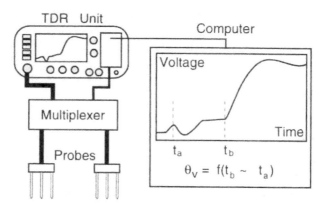

a) CSIRO's PyeLab system

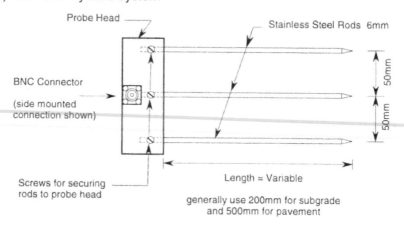

b) CSIRO's TDR moisture probe.

Figure 3-1. CSIRO's TDR soil moisture system and probe; an example of a three-rod configuration (from Baran, 1994).

influence: 1) what is the diameter of a cylinder effectively sampled around the probe when inserted vertically into the soil, and 2) if the probe is installed horizontally in the soil, how close can it be to the surface? Findings from various studies are summarized in Table 3.1.

In addition to the rod spacing, probe performance is affected by rod diameter. Knight (1992) presented a theoretical investigation of the size of the cylinder of influence. He recommended that

Table 3.1. Effective sampling volume of TDR probes.

Rod spacing, s (mm)	Vertical installation - cylinder radius	Horizontal installation - minimum distance to ground surface	Reference
2-prong			
10		1.00s	Petersen (1995)
20		0.75s	Petersen (1995)
25	1.00s		DeClerck (1985)
50	0.60s	0.40s	Baker and Lascano (1989)
50	0.70s		Topp and Davis (1985)
50		0.40s	Petersen (1995)
	0.5(s+d)	0.5(s+d) d = rod diam.	Knight et al. (1994)
3-prong			
	0.70s	0.70s	Knight et al. (1994)

probes be designed with a ratio of rod diameter to spacing (d/s) greater than 0.1 in order to prevent too much energy being concentrated closely around the rods.   Working with probes appropriate for laboratory scale testing, Petersen (1995) found that the smaller the rod diameter, the higher the resolution and the closer measurements can be made to the soil surface. Good determination of water content was found with d/s as small as 0.02.   For rod spacings of 10, 20, and 50 mm, measurements were accurately made as close to the soil surface as 10, 15, and 20 mm (Table 3.1). As a general rule of thumb, for three-prong TDR probes, the center rod diameter should be no smaller than 10 times the mean particle size of the soil or porous material being investigated (Knight, 1992; Zegelin et al., 1989).

A brief summary of prototype and commercially available probe designs is provided in Table 3.2. When selecting probes for field studies, the two-prong designs have the important advantage of a larger sampling volume as shown in Table 3.1. The use of electrodes which are small in diameter and closely spaced should be avoided to minimize the errors that may occur as a result of preferential shrinkage away from the rods (Whalley, 1993). If a short probe is required, then the three-prong probe has the advantage of producing a much clearer signal that can be interpreted more easily (Zegelin et al., 1989). Furthermore, the three-prong probe eliminates the need for balancing transformers (baluns) as discussed in Chapter 10.

SITE-SPECIFIC CALIBRATION METHODS

TDR measurement of soil moisture is based on empirical correlations between volumetric water content and the travel time of a voltage pulse along a probe embedded in soil. Depending upon the level of accuracy desired, published correlations can be employed or site-specific calibrations can be performed. A variety of procedures have been used to develop these correlations and a few will be illustrated below by examples.

Logsdon (1994) obtained undisturbed soil samples by using hydraulic pressure to insert steel stove pipe (100 mm inner diameter, 600 mm long) horizontally into the sides of pits. The usable samples had soil lengths ranging from 320 to 400 mm within the stovepipe column. Samples were collected in pit walls at depths of 0.22 m, 0.46 m, and 1.11 m. In the laboratory, two-rod 200-mm-long stainless steel probes, attached to a 1:1 balun as described by Spaans and Baker (1993) and discussed in Chapter 10, were inserted into the wet soil samples. The samples were oriented horizontally during the course of the calibration procedure. Coaxial cables (50 ohm) connected the probes to a multiplexer and the multiplexer to a cable tester. The cable tester was controlled by a datalogger which converted the distance between the signal pulse

Table 3.2. Probes used for moisture monitoring.

Rod diameter, d (mm)	Rod spacing, s (mm)	Diameter/ spacing, d/s	Rod length, l_p (mm)	Reference
			2-prong	
9.5	50.8	0.19	200, 300, 400, 600	Soilmoisture Equipment Co. (1995)
6.0	40.0	0.15	160	IMKO GmbH (1995)
4.0	25.0	0.16	117	Hook et al. (1992)
3.5	20.0	0.18	110	IMKO GmbH (1995)
3.2	30.0	0.11	300	Campbell Scientific, Inc. (1992a)
2.0	16.0	0.13	100	Easy Test (1994)
			3 prong	
10.0	65.0	0.15	500	van Schelt et al. (1994)
9.5	38.1	0.25	203	Rada et al. (1994)
8.0	35.0	0.23	160	IMKO GmbH (1995)
6.0	50.0	0.12	200, 500	CSIRO type; Baran (1994); Look et al. (1994a)
6.0	50.0	0.12	300, 500	QT type; Baran (1994); Look et al. (1994a)
4.8	44.7	0.11	300	Campbell Scientific, Inc (1992a)
4.8	25.0	0.19	100, 150	Knowlton et al. (1994)
4.7	30.0	0.16	150	Zegelin et al. (1989)
3.2	30.0	0.11	200	Vadose Zone (1994)
3.2	25.0	0.13	203	Kotdawala et al. (1994)
3.0	25.0	0.12	200	Soilmoisture Equipment Co (1995)

reflections to the inverse of propagation velocity, $1/V_p$.

Samples were weighed and TDR readings taken periodically over a 3- to 4-week drying phase. Because of soil instability, samples were kept in the stovepipes during this phase. Gravimetric water contents were converted to volumetric water contents by multiplying by the bulk density (Equation 3-4). Four TDR waveform results were averaged for each measurement of each sample. Drying occurred at either end of the exposed column so water content was not uniform over the length of the column. This nonuniform drying pattern could allow gaps to develop, which would reduce measured $1/V_p$ (Annan, 1977; Torres et al., 1992). For some columns that were not initially wet enough, water was misted on the ends periodically over several days before TDR readings were taken.

Calibration curves expressed volumetric water content (θ_v) as a linear function of $1/V_p$ similar to procedures used by Ledieu et al. (1986) and Herkelrath et al. (1991). The scatter of data points suggested that there was no benefit in using a more complicated calibration formula. Calibration curves were grouped by slope to express general calibration for a variety of soils. When necessary, bulk density was included as a variable in the calibration-regression equations. See the discussion of soil density influence later in this chapter.

In another study (Zegelin and White, 1994), coal samples were prepared at various water contents. Coal was placed in polyethylene bags, water was added, and then the samples were sealed and intimately mixed. Samples were allowed to stand for 24 h to equilibrate. Measurements showed that for most coal samples, equilibration was rapid and was attained in less than one hour (i.e., TDR readings did not change). Known weights of the sample were packed to a known bulk density in 1-liter polyvinyl chloride (PVC) containers. The apparent dielectric constant of the mixture, K_a, defined in Equation 2-33, was determined using a three-rod TDR probe of length 150 mm, rod diameter 4.7 mm, and rod spacing 25 mm. Pulse travel time in the TDR probe was determined with the PyeLab TDR system, shown in Figure 3-1a, and results were automatically analyzed and logged. Ten measurements were made on each sample. The volumetric water content of the sample was also determined by measuring the sample bulk density and then drying at 105°C in a forced convection oven under a nitrogen atmosphere for 12 to 24 h.

TOPP'S CALIBRATION EQUATION–THE "YARDSTICK"

The high dielectric constant of water ($K_{water} \approx 80$) imparts a very strong dependence of a moist soil's bulk dielectric constant on water content (Topp et al., 1994). Following on with the early work of Davis and Chudobiak (1975), an empirical relationship between dielectric constant and volumetric water content was determined in the laboratory for a wide range of soils, using coaxial sample holders. This empirical approach made no assumption about the state of water in porous materials. Instead, it related the TDR-measured apparent dielectric constant, K_a, to volumetric water content based on oven drying to 105°C. Polynomial regression equations were used to represent the relationship for different soils. Topp et al. (1980) showed for a wide range of mineral content soils that a single equation was adequate and was practically independent of soil bulk density, ambient temperature, and salt content. Their relation is now widely used as a calibration curve and is referred to as Topp's equation,

$$\theta_v = 4.3x10^{-6}K_a^3 - 5.5x10^{-4}K_a^2 + 2.92x10^{-2}K_a - 5.3x10^{-2}. \quad (3-7)$$

In a surprising number of instances this equation has been shown to be quite broadly applicable, even for gravelly soil (Dalton, 1992; Dalton et al., 1984; Dalton and van Genuchten, 1986; Dasberg and Dalton, 1985; Drungil et al., 1989; Heimovaara, 1994; Heimovaara and Bouten, 1990; Roth et al., 1992; Skaling, 1992; Topp et al. 1984; Topp and Davis, 1985; Van Wesenbeek and Kachanoski, 1988; Whalley, 1993; Zegelin et al., 1989; Zegelin et al., 1992). This led to the use of the term "universal" for this equation with the appropriate caveat that "in organic soils or heavy clay soils problems arise which may require site-specific calibrations,..." (Zegelin et al., 1992). The successful use of TDR depends upon ensuring that the calibration equation used is appropriate for the soil and application. Equation 3-7 and comparable calibration equations are applicable where changes in water content are desired rather than determination of absolute values.

CALIBRATION COEFFICIENTS FOR GENERAL RELATIONSHIPS

To appreciate reasons why material type and calibration procedures can cause deviations from Equation 3-7, consider a study (Zegelin and White, 1994) using coal samples from various mines in Australia. Coals are complex materials consisting of a plethora of organic and inorganic constituents and possessing intricate micro and macro pore structures. The physical properties of coal vary widely from the colloidal brown coals with low carbon contents to rigid, graphite-like structures in the high-carbon anthracite. Interaction of water with coal varies markedly with carbon content (or rank). Water can exist in coals in a wide range of configurations, from tightly bound monomolecular layers (Lynch and Webster, 1982) to pockets of bulk water in larger pores (Van Krevelin, 1981).

The dielectric constant of dried coal containing 70% to 80% carbon depends markedly upon how the coal was dried (Guitini et al., 1987). For oven-dried coals, the dielectric constant varies between 3.5 and 5 for carbon contents between 70% and 90%. This is also the range for air-dried coals with carbon contents between 80% and 90%. Air-dried coals with carbon content between 70% and 80% may have a dielectric constant between 5 and 13. The dielectric constant of air-dried coal appears to increase rapidly with carbon content above 90%. Clearly, for coals below 80% carbon, the colloid content is high and the amount of bound water appears significant. Zegelin and White (1994) proposed that the TDR technique will work best for coals with carbon contents between 80% and 90% where the dielectric constant of dry coal is normally about 3.8 (Balanais et al., 1978). The coefficients of the third degree polynomial regressions are shown in Table 3.3 along

Table 3.3. Calibration coefficients (Zegelin and White, 1994).

Material	A (10⁻⁵)	B (10⁻³)	C (10⁻²)	D (10⁻²)	r²
Light textured soils Equation 3-7	0.4300	-0.5500	2.92	-5.3000	
Liddell coal	8.0200	-3.1950	5.367	-7.5550	0.997
Bayswater coal	0.7226	-0.7585	2.902	-11.6700	0.993
Brown coal (average)	3.2850	-2.4810	6.554	0.4228	0.996
Nickel ore	11.6000	-5.4840	9.251	-20.9100	0.983

with the correlation coefficient, r^2, for these fits.

The particle size of coal was found to have little effect on the $\theta_v(K_a)$ calibration curve of a crushed subbituminous coal from the Liddell colliery for different size fractions (Zegelin and White, 1994). Whole coal refers to the industrial sample passing through a 19 mm sieve. Although there was scatter in the data, the calibration curve appeared to be independent of the particle size for this size fraction. For larger particle sizes, the probe geometry will need to be adjusted to ensure the sample volume of the TDR measurement is representative, as mentioned in the discussion of probe design. To assess the effect of carbon content, Zegelin and White (1994) examined a meta-anthracite sample from the Bayswater colliery, NSW, Australia. This sample had high electrical conductivity that greatly attenuated the TDR trace. The calibration curve had a slope about 50% greater than that of the Liddell sample, which indicates the apparent dielectric constant of water in the meta-anthracite is greater than that of bulk water. They also found that the TDR signal attenuation in coke samples was so severe that the TDR technique was not applicable with uninsulated three-rod probes. For the more graphite-like materials, they suggest that there is also an appreciable contribution of the imaginary component of dielectric constant, due to electrical conductivity. See the discussion of the influence of frequency later in this chapter.

DIELECTRIC PROPERTIES AND ABSOLUTE WATER CONTENT

The dielectric constant of a material ($K=\epsilon/\epsilon_o$) can be defined as the ratio of the electrostatic capacity of condenser plates, separated by the material, to the capacity of the same condenser with vacuum between the plates (Mitchell, 1976). It is a measure of the ease with which molecules can be polarized and oriented in an electric field. Water at 20°C has a dielectric constant of about 80, while mineral soil grains can have a dielectric constant less than 5.

The dielectric properties of soil are a function of bulk density, specific surface (Dirksen and Dasberg, 1993), particle shape (Sivhola and Lindell, 1988), and organic content (Hilhorst and Dirksen, 1994). At present, there is no universal relationship between dielectric constant and soil water content for precise absolute measurements. Therefore, calibration of dielectric sensors for soils of interest is unavoidable, particularly if high accuracy ($\pm$ 0.01 m^3/m^3) is required (Petersen, 1995) or the frequency content of the TDR pulse is varied.

Influence of Soil Type

Bohl and Roth (1994) conducted an investigation based in part on a database collected jointly by the Technical University of Berlin, Germany, and the Institute of Agrophysics, Lublin, Poland (Malicki et al., 1994; Roth et al., 1992). For the model tests, all mineral and organic soils from this database were used. A second data set for investigating the influence of bound water was obtained from a range of sand–peat mixtures. Relevant characteristics of the soils and the mixtures are listed in Tables 3.4 and 3.5. The investigations were based on a total of 418 triplet determinations of apparent dielectric constant, K_a, volumetric water content, θ_v, and porosity, η, in 40 different soil horizons and mixtures. The plot of apparent dielectric constant versus volumetric water content of all measurements in Figure 3-2 shows that the relation between θ_v and K_a varies over a wide range.

In a dry clay, adsorbed cations are tightly held by the negatively charged clay surfaces. Cations in excess of those needed

Table 3.4. Description of mineral soils (Bohl and Roth, 1994).

Location	Family and Order	Horizon	Texture (wt%)			Porosity	Organic carbon (wt%)
			Clay	Silt	Sand		
Brazil							
Paraná	Haplic ferralsol	Ah	12	6	82	0.450	1.12
		Bws	18	8	74	0.464	0.35
Germany							
Grunewald	Haplic luvisol	Bt	23	75	2	0.435	0.19
	Haplic podsol	Bhs	2	10	88	0.449	1.93
Grunewald	Arenic cambisol	Aeh	2	14	84	0.592	5.91
		Bsw	1	9	90	0.427	1.50
		Bw	0	2	98	0.380	0.67
		Cv	0	2	98	0.369	0
Italy							
Elba	Chromic luvisol	Ah	23	31	46	0.445	1.00
		Bt	34	31	35	0.419	0.21
	Luvic calcisol	Ah	24	26	50	0.428	1.17
		Btk	46	28	26	0.482	0.63

Table 3.4. (continued)

Location	Family and Order	Horizon	Texture (wt%) Clay	Texture (wt%) Silt	Texture (wt%) Sand	Porosity	Organic carbon (wt%)
Poland							
Werbkovice	Haplic phaeozem	Ap	46	52	2	0.447	3.36
Jablon	Mollic gleysol	Ah	40	54	6	0.525	1.89
Rudnik	Eutric cambisol	Bt	58	34	8	0.540	0.24
Rogòzno	Orthic podsol	Bhs	9	5	86	0.400	0.09
Tarnawatka	Eutric cambisol	Bt	35	63	2	0.566	0.48
Grabowiec	Eutric cambisol	Bt	48	47	5	0.570	1.08
Switzerland							
Winzlerboden	Luvic arenosol	Ah	7	9	84	0.437	3.30
Buchberg	Gleyic luvisol	Bt	14	49	37	0.450	0.90
Abist	Gleyic cambisol	Ah	20	31	49	0.511	1.27

Table 3.5. Description of organic soils, Berlin, Germany
(Bohl and Roth, 1994).

Soil Type Family and Order	Horizon	Porosity (cm^3/cm^3)	Organic carbon (wt%)
Arenic cambisol	Ofh	0.619	37.6
Terric histosol	H1	0.920	32.8
Fibric histosol	H2	0.723	41.3
	H1	0.854	43.2
	H2	0.896	47.1
	H	0.785	16.2
	H1	0.785	54.8
	H2	0.580	48.7
Cambic arenosol	O1	0.680	44.0
	O2	0.714	41.8
	O3	0.660	38.9
	O4	0.648	32.2
	Ah	0.528	10.5

to neutralize the electronegativity of the clay particles and their associated anions are present as salt precipitates (Mitchell, 1976). When the clay is placed in water, the precipitated salts go into solution. Because the adsorbed cations are more highly concentrated near the surfaces of particles, there is a tendency for them to diffuse away in order to equalize concentrations throughout. Their freedom to do so, however, is restricted by the negative electric field originating in the particle surfaces. The escaping tendency due to diffusion and the opposing electrostatic attraction leads to a cation distribution adjacent to a clay particle in suspension as shown in Figure 3-3. The distribution of cations adjacent to a clay is analogous to that of the air in the atmosphere, where the escaping

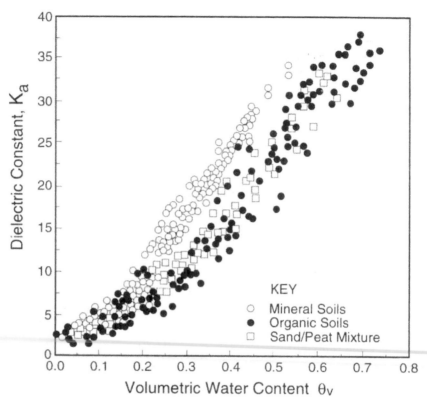

Figure 3-2. Influence of soil type on the correlation of water content and the apparent dielectric constant (from Bohl and Roth, 1994).

tendency of the gas is countered by the gravitational attraction of the earth. Anions, however, are repelled by the negative force fields of the particles, giving the distribution shown in Figure 3-3. This phenomenon is sometimes termed "negatively charged" (Marshall, 1964). The negative surface and the distributed charge in the adjacent phase are together termed the diffuse double layer, and it affects the ease with which water molecules can align themselves in a changing electric field.

Organic particles may be strongly adsorbed on mineral surfaces, and this adsorption modifies both the properties of the minerals and the organic material itself. The adsorption of organic matter on mineral surface and edges may lead to interparticle negative adsorption.

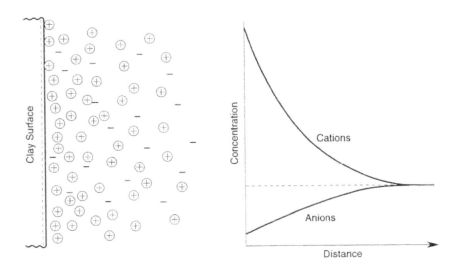

Figure 3-3. Distribution of ions adjacent to a clay surface according to the concept of the diffuse double layer (Mitchell, 1976; reprinted by permission of John Wiley & Sons, Inc.).

       Organic matter in soils is complex both chemically and physically and varies with age and origin (Mitchell, 1976). It may occur in any of five groups: carbohydrates; proteins; fats, resins, and waxes; hydrocarbons; and carbon. The humic faction is gel-like in properties and bonding or coating of large particles affects the interaction between soil particles and soil water. This complex interaction affects the ease with which water molecules can align themselves in a changing electric field. See the discussion of frequency influence and the discussion of alternative calibration equations later in this chapter.

Influence of Soil Density and Shrinkage

Logsdon (1994) studied twenty-two samples of which nine coincided with one calibration equation as shown in Figure 3-4a. Bulk densities for these samples ranged from 1.3 to 1.5 Mg/m³, clay from 20% to 35%, and sand from 40% to 60%. Excluding the Nicollet 0.46 m, the regression equation was

$$\theta_v = -0.0928 + 0.0964/V_p. \qquad (3\text{-}8)$$

The resulting calibration has a smaller slope than Topp's relationship (Equation 3-7). There were also five rather dissimilar samples grouped together that had either low (15%) or high (32% to 45%) clay content, and bulk density, ρ_b ranging from 1.35 to 1.65 Mg/m³ as shown in Figure 3-4b. A different calibration was developed for this group to account for the influence of bulk density,

$$\theta_v = -0.6709 + 0.1288/V_p + 0.2690\rho_b. \qquad (3\text{-}9)$$

Ledieu et al. (1986), Herkelrath et al. (1991) and Whalley (1993) theoretically determined that there should be a linear relation between travel time and θ_v. Since travel time is linearly related to $1/V_p$, the empirical linear relations between $1/V_p$ and θ_v in Equations 3-8 and 3-9 are consistent with this theory. Equation 3-9 is also consistent with Whalley's finding that the intercept is determined by bulk density. For all of the calibration samples, $1/V_p$ measurements at θ_v less than 0.20 to 0.25 m³/m³ were much smaller and the $\theta_v(1/V_p)$ relation was not linear. This difference could be related to the drying gradient which not only caused heterogeneities of θ_v along the probes (Dasberg and Hopmans, 1992; Roth et al., 1992) but also reduced the values of measured θ_v. Gaps could have developed between the probe rods and soil due to soil shrinkage, but none were observed during removal of the probes. Gaps as small as 1 mm will reduce the measured dielectric constant (Annan, 1977; Whalley, 1993).

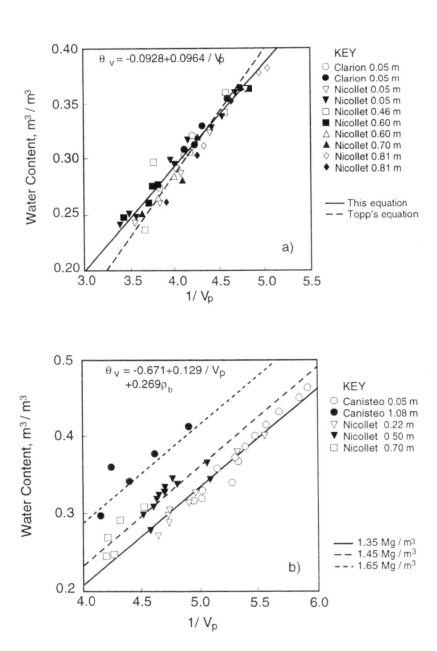

Figure 3-4. Calibration of TDR for θ_v: (a) as a function of $1/V_p$; (b) as a function of $1/V_p$ and ρ_b; the units of $1/V_p$ are nanoseconds per foot (from Logsdon, 1994).

Malicki et al. (1994) studied 894 quadruplets of K_a, θ_v, ρ_b, and ρ_d which were made with sixty-two samples composed of: 1) mineral soils having different textures, bulk densities, and particles densities, 2) organic soils having different amounts of organic matter, bulk densities, and dry densities, 3) standard pot soils having different textures, bulk densities, and dry densities, 4) artificial peat-loess and peat-sand mixtures having different bulk densities, and dry densities, 5) marine and alluvial sands, 6) forest litter, and 7) wood bark, shavings, and sawdust. They developed the following empirical relationship

$$\theta_v = [K_a - 0.819 - 0.168\rho_b - 0.159\rho_b^2] \, / \, [7.17 + 1.18\rho_b]. \quad (3\text{-}10)$$

Deviations between TDR and gravimetrically measured water contents were analyzed by comparison with a Gaussian normal distribution using the chi-squared test. The results indicate that a hypothesis of a random density-corrected soil moisture error cannot be rejected. This implies (for the materials investigated) that the influence of matrix properties other than bulk density can be neglected if a standard error in θ_v of ± 0.03 m³/m³ is acceptable.

Influence of Temperature

Researchers at the Swedish University of Agricultural Sciences have used the TDR technique for estimation of water content in soils ranging from sand to heavy clay over a range of temperature and moisture regimes (Alvenäs and Stenberg, 1995). However, they have experienced several problems when evaluating the data. They noticed a dependence of the estimated soil water content on temperature. In particular, the daily oscillation in estimated soil water content has been found to be synchronized with daily temperature oscillation. The difference between the highest and lowest water contents during a clear day could be $\Delta\theta_v = 0.05$ m³/m³.

The influence of temperature on soil moisture measurement has been evaluated by several researchers, as shown in Table 3.6, but there has not been a consensus. It is a function of the soil type and water content itself. The dielectric constant of pure water decreases as temperature increases as shown in Figure 3-5. The following tabulation for water shows the nature of this variation (Mitchell, 1976):

Temperature, T		Dielectric Constant of water, K_{water}	$K_{water}T$
(°C)	(°K)		(°K)
0	273	88	2.40×10^4
20	293	80	2.34×10^4
25	298	78.5	2.34×10^4
60	333	66	2.20×10^4

The small variation of the product $K_{water}T$ with change in temperature means, theoretically, that the influence should not be great. This assumes, of course, that the dielectric constant is unaffected by the particle surface forces and ionic concentration. It also accounts, in part, for apparently contradictory findings reported in the literature on the effects of temperature change on TDR measurement of dielectric constant and water content. Temperature dependence from several sources is summarized in Table 3.6 to illustrate the variation and order of magnitude. If the behavior was consistent with that of pure water, then the change in apparent dielectric constant, ΔK_a, and water content, $\Delta \theta_v$, would be negative as temperature increases. However, as shown by the table, there is no consistent trend even with regard to sign.

    Some insight can be gained from the study by van Loon et al. (1995) with carbohydrate solutions. They found that temperature has a complex influence on the measured dielectric constant: at high water contents the relation is negative, while at low water contents this relation is positive. This change in sign results in a cross-over point at a water content $\theta_v \approx 0.55$. While this result may not be directly applicable to soils, it demonstrates that the influence of temperature may also be a function of water content itself.

    From a practical point of view, it is very important to note the order of magnitude. Topp et al. (1980) found that the TDR-measured dielectric constant varied from 18 to 20 as the temperature increased from 10°C to 36°C which gives $\Delta K_a/\Delta T = 0.077°C^{-1}$.

Table 3.6. Influence of temperature on TDR measured dielectric constant and volumetric water content.

Reference	Soil or solution	Temp. T (°C)	TDR K_a^+	TDR $\Delta K_a/\Delta T$ (°C⁻¹)	Volumetric θ_v^{++} (m³/m)	Volumetric $\Delta\theta_v/\Delta T$ (°C⁻¹)
Mitchell (1976)	water	20 60	80 66	-0.3500		
Topp et al. (1980)	clay loam $\theta_v = 0.32$	10 36	18 20	0.0769		0.0015
Halbertsma et al. (1995)	sand and loam	5 70			0.350 0.300	-0.0008
	sand	5 70			0.006 0.006	0
	clay	5 70			0.492 0.492	0
Verstricht et al. (1994)	Boom clay $\theta_v = 0.39$	10 40	29 34	0.1667		0.0007

+Apparent dielectric constant   ++Water content

Table 3.6. (continued)

Reference	Soil or solution	Temp. T (°C)	TDR K_a^+	TDR $\Delta K_a/\Delta T$ (°C^{-1})	Volumetric θ_v^{++} (m^3/m)	Volumetric $\Delta\theta_v/\Delta T$ (°C^{-1})
Alvenäs and Stenberg (1995)	heavy clay 4 cm^{+++}	12 30			0.230 0.270	0.0022
	20 cm^{+++}	16 20			0.340 0.340	0
van Loon et al. (1995)	saccharose $\theta_v = 0.4$	5 40	27 40	0.3714		
	$\theta_v = 0.6$	5 40	57 57	0		
	$\theta_v = 0.9$	5 40	79 75	-0.1143		

+Apparent dielectric constant   ++Water content   +++Depth

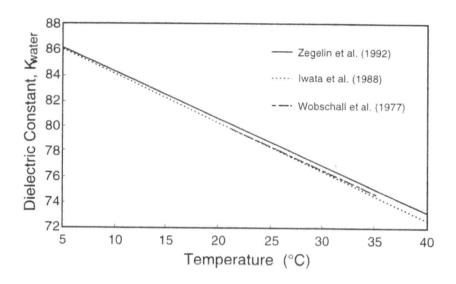

Figure 3-5. Dielectric constant of water declines as temperature increases.

In Table 3.6 this has been converted to a change in water content per degree $\Delta\theta_v/\Delta T = 0.0015°C^{-1}$ using the calibration curve in their paper. They concluded that, for practical purposes, there is no influence of temperature on TDR measurement of soil moisture. This judgement must be made by each user based on whether they are attempting to acquire absolute or relative values of soil moisture, as well as the desired sensitivity of the measurement.

Frequency Considerations

Dielectric properties of a material depend upon the ease with which its constituent molecules can be polarized (Mitchell, 1976). There are two contributions to the polarization that develop in an electrical field: distortion polarization, which is instantaneous, and orientation polarization, which is time-dependent. In a low-frequency electromagnetic field, dipoles can rotate with the changes in current direction. As the frequency increases, however, less time is available for dipole rotation. Above some frequency, the molecules will no longer be able to follow the field and the dielectric constant will drop. The dielectric losses and the frequencies at which they occur are dependent on the intermolecular bond types and strengths, which are, in turn, a function of soil mineralogy and organic content. Hence the results of dielectric constant measurements have been used to infer details of water–soil structure. TDR measurement of soil moisture is similar in principle to the induced-polarization method applied in mineral exploration (Yong and Hoppe, 1989). The main difference stems from the necessity of detecting high-frequency polarization phenomena occurring in the pore fluid for moisture content determination, as opposed to low-frequency interfacial polarization indicative of mineral content.

The dielectric properties of materials (Hasted, 1973), such as soils, can be described by a complex representation of the dielectric permittivity ($\epsilon = K\epsilon_o$),

$$\epsilon = \epsilon' - j\epsilon'' \ , \qquad\qquad (3\text{-}11)$$

where the real part of the permittivity, ϵ', is a measure of the polarizability of the material constituents, including water (Hilhorst and Dirksen, 1994). The imaginary part of the permittivity, ϵ'', represents the energy absorption. The energy is absorbed by dielectric losses and ionic conduction,

$$\epsilon'' = \epsilon''_d + \sigma_{dc}/(\epsilon_0 \omega) \ , \qquad\qquad (3\text{-}12)$$

where ϵ_d'' is the dielectric loss, σ_{dc} is the dc or static ionic conductivity, ϵ_o is the dielectric constant of a vacuum, and $\omega \ (= 2\pi f)$ is angular frequency.

The reorientation of polar molecules, such as water, in an electric field, although rapid, takes some time. With increasing frequency, the molecules become less able to follow the increasingly rapid alternating electric field. At these higher frequencies, ϵ' will fall to much lower values as shown in Figure 3-6. The frequency at which ϵ' has decreased to half its lower frequency value is called the relaxation frequency. At this frequency the dielectric energy loss, ϵ_d'', has its maximum. At sufficiently low frequencies, however, dielectric losses are negligible compared with the ionic conductivity (i.e., $\epsilon_d'' << \sigma_{dc} / \epsilon_o \omega$).

The dielectric constant ($K = \epsilon/\epsilon_0$) of the soil components in the TDR frequency range of 1 to 1,000 MHz is about 80 for water, 3 to 4 for major soil minerals, and 1 for air. For dry soil, the real part of the dielectric constant, ϵ', is less than 5. For free water, ϵ' $\approx$ 80 and the relaxation frequency is 17,000 MHz (Hilhorst and Dirksen, 1994). For soil-bound water, ϵ' varies between about 4 and 80, with a relaxation frequency of less than 1,000 MHz, depending upon the binding forces. Therefore, the real part of the permittivity of wet soil is dominated by the volumetric water content (i.e., $\epsilon' = f(\theta_v)$).

TDR is a wide-frequency bandwidth technique. The ability to resolve or measure the water content within short lengths of transmission lines improves as the frequency bandwidth increases (Topp and Davis, 1985). Frequencies above about 1,000 MHZ are attenuated very rapidly by soils. Therefore, the ideal TDR pulser/sampler has as large a bandwidth as possible below 1,000 MHz. The TDR signal attenuation in coarse-grained soil is much less than in fine-grained soils in which the upper frequency limit may be 20 MHz. In heavy clay soil at high water content, the worst case, the upper frequency limit may be less than 1 MHz. This effect is indicated by the horizontal dashed lines in Figure 3-6.

Ionic double layers associated with colloidal soil particles (clay) result in high (>80) apparent dielectric constants for these soils at frequencies below 10 MHz. This effect is often referred to as the Maxwell-Wagner effect (Wagner, 1914). The effect of ionic conductivity in the absence of dielectric losses, $\epsilon''(\sigma)$, is shown in Figure 3-6 with the dotted line (Hilhorst and Dirksen, 1994). This suggests that σ dominates ϵ'' at lower frequencies (i.e., σ_{dc}), but White et al. (1994) suggest that conductivity dominates at high frequencies. They tested a three-phase effective medium model with graphite–sand mixtures in which conductivity could be varied systematically. Measured calibration relations were linear, as

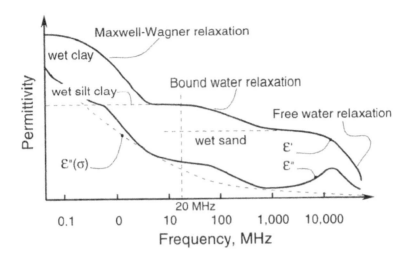

Figure 3-6. Qualitative representation of dielectric properties of wet soils as function of frequency (Hilhorst and Dirksen, 1994).

predicted by the model, but the $K_a(\theta_v)$ calibrations had slopes up to 250% larger than expected from the bulk phase properties of water. Independently determined low-frequency electrical conductivities of the samples were insufficient to account for the increases in slopes. Additions of electrolyte solution to samples confirmed that low-frequency conduction losses were not the principal cause of $K_a(\theta_v)$ slope increases. The model suggests that the water phase of these materials behaves as if its dielectric properties were substantially different from bulk water. It is suggested, by analogy with insulator–conductor composites, that high frequency conductivity losses may account for slope increases in porous materials.

ALTERNATIVE CALIBRATION EQUATIONS FOR SOIL MOISTURE

The widespread use of TDR has resulted in a number of applications in which Topp's calibration (Equation 3-7) cannot be used, and some effort has been devoted to finding alternative relationships between apparent dielectric constant and water content (Dirksen and Dasberg, 1993; Herkelrath et al., 1991; Paquet et al., 1993;

Pepin et al., 1992; Roth et al., 1992). The need for alternatives to Equation 3-7 has been motivated by attempts to use TDR in materials with higher surface area, higher electrical conductivity, and/or bulk density outside the range of those soils from which it was determined (Topp et al., 1994). Roth et al. (1992) have shown that mineral soils having bulk densities up to 1.55 Mg/m^3 were reasonably well represented by Equation 3-7. Dirksen and Dasberg (1993) selected mineral soils having a wider range of clay content and bulk density to show conditions under which this calibration does not provide an adequate representation of the dielectric response. As summarized by Zegelin et al. (1992; p. 206): Topp's calibration "works best in coarser textured soils, but has problems in fine textured, dense, heavy clay soils." If a high degree of accuracy is required, it is advisable to test its applicability and construct a site-specific calibration, if necessary.

A number of recent articles report TDR calibrations for organic soil (Herkelrath et al., 1991; Paquet et al., 1993; Pepin et al., 1992; Roth et al., 1992). These results are beginning to show that a fairly consistent relationship exists for organic soil, but it is not universal. For organic soil, the relationship has a lower slope at low water contents than it does for most mineral soils. This is believed to arise from the large wettable surface area in organic soils.

An alternative to the use of empirically derived calibrations is to make use of dielectric mixing formulae (Dirksen and Dasberg, 1993; Roth et al., 1990). In the mixing law approach, the soil is considered a mixture of three phases: water, soil, and air. The three-phase mixing law gives (Roth et al., 1990)

$$K_a = [\theta_v K_{water}^{\beta} + (1 - \eta) K_s^{\beta} + (\eta - \theta_v) K_{air}^{\beta}]^{1/\beta} \qquad (3\text{-}13)$$

where K_a is the apparent dielectric constant of the composite mixture; η is the soil porosity; K_{water}, K_s, and K_{air} are dielectric constants of water, soil solids, and air, respectively; and β is a geometric factor that depends upon the spatial arrangement of the mixture and its orientation in the electric field. Having values between +1 and -1, β has been optimized to give the closest relationship to measured data. In most cases, β has a value about 0.5. The parameter can also account for any contributions from frequency-dependent complex dielectric permittivities of the soil components. This approach assumed that there is no interaction between water and soil.

In a similar approach using mixing laws, a bound water phase is added in which the amount of bound water or its dielectric constant is adjusted to match the model to measured data. This four-phase model distinguishes between a free water phase (θ_{fw} and K_{fw}) and a bound water phase (θ_{bw} and K_{bw}) (Dirksen and Dasberg, 1993; Dobson et al., 1985; Bohl and Roth, 1994)

$$K_a = [\ \theta_{bw}K_{bw}^{\beta} + \theta_{fw}K_{fw}^{\beta} + (1-\eta)K_s^{\beta} + (\eta-\theta_v)K_{air}^{\beta}\ ]^{1/\beta}. \quad (3\text{-}14)$$

The Maxwell-De Loor model (De Loor, 1968) also distinguishes these four phases,

$$K_a = \frac{3K_s + 2(\theta_v - \theta_{bw})(K_{fw} - K_s) + 2\theta_{bw}(K_{bw} - K_s) + 2(\eta - \theta_v)(K_{air} - K_s)}{3 + (\theta_v - \theta_{bw})(K_s/K_{fw} - 1) + \theta_{bw}(K_s/K_{bw} - 1) + (\eta - \theta_v)(K_s/K_{air} - 1)} \cdot (3\text{-}15)$$

These mixing laws go to the correct limiting values for dry soil and for completely water-filled conditions, and the fitting parameters have been adjusted successfully to produce physically valid equations (Dirksen and Dasberg, 1993). In contrast to applicable empirical equations that do not require prior knowledge of the soil properties, the application and use of mixing formulae require varying levels of information about the soil components: values for dielectric constants (K_{water}, K_s, and K_{air}), and porosity or bulk density. Alharthi and Lange (1987) have shown that substitution of known values for the dielectric constants and setting at $\beta = 0.5$ in Equation 3-13 gives a simple expression

$$\sqrt{K_a} = 7.83\theta_v + 1.594. \quad (3\text{-}16)$$

This linear relationship represents a significant improvement over the polynomial calibration curves because it has only two parameters and, being linear, is much easier to use where applicable, such as for coarse and medium textured soils.

Bohl and Roth (1994) found that the Maxwell-De Loor model (Equation 3-15) yielded the most accurate $\theta_v(K_a)$ relation for mineral soils and for organic soils as shown by the comparison in Table 3.7. The optimal value for the volumetric fraction of the bound water was found to be $\theta_{bw} = 0.03$ in mineral soils and $\theta_{bw} = 0.06$ in organic soils. In all cases, they used $K_{bw} = 3.2$ for the dielectric constant of the bound water and $K_{fw} = 80.36$ for free water. The dielectric constant of the soil matrix, K_s, was assumed as 3.9 in mineral soils and as 5 in organic soils.

Accuracy of the water content determined from the mixing models is comparable to that obtained from empirical calibrations as shown in Table 3.7, although models are based on highly simplifying assumptions and the dielectric constants for the soil constituents have been assumed to be identical for all soils. Despite this encouraging result, which has been confirmed for undisturbed samples, there remain some open questions particularly regarding the physical relevance of Equation 3-15 at high water contents and the origin of the pronounced trend present in the residuals of all models.

Table 3.7. Parameters and accuracy of the applied models and empirical approaches (Bohl and Roth, 1994).

$\theta_v(K_a)$ relation	θ_{bw} optimized[*] (cm³/cm³)	Standard error of θ_v (cm³/cm³)
Mineral (n = 263)		
β 3-phase Equation (3-12)		0.027
β 4-phase Equation (3-13)	-0.005 ± 0.002	0.026
Maxwell-De Loor Equation (3-14)	+0.030 ± 0.002	0.019
Topp et al. (1980) Equation (3-7)		0.029
Malicki et al. (1994)		0.017
Organic (n = 155)		
β 3-phase Equation (3-12)		0.065
β 4-phase Equation (3-13)	+0.065 ± 0.002	0.028
Maxwell-De Loor Equation (3-14)	+0.054 ± 0.003	0.030
Roth et al. (1990)		0.039
Malicki et al. (1994)		0.033

[*] +/- indicates one standard deviation

Chapter 4
FIELD EXPERIENCE AND VERIFICATION OF SOIL MOISTURE MEASUREMENT

The viability of TDR to measure water content of soils must be assessed by comparison with other techniques commonly employed. This chapter presents such comparative studies with lysimeters and Bowen's ratio, neutron probes, and nuclear gauges. For the most part, these comparisons have been made with partially saturated soils. After presentation of typical agricultural field validation studies, additional case studies are presented to demonstrate use of TDR to measure soil moisture associated with landfill covers, compacted embankments, and pavement subgrade.

LYSIMETER AND BOWEN RATIO

In a wheat field with a thick clay B horizon, Zegelin et al. (1992) carried out comparative tests between TDR measured changes in soil-water and those found from a weighing lysimeter and the Bowen ratio technique. The trends in stored water changes measured by TDR and lysimeter are similar, where the average absolute deviation of TDR measurements from lysimeter values was 1.4 and 1.6 mm, respectively, an average deviation of less than 10% in both cases. TDR is advantageous as it offers the possibility of easily sampling water content at hourly intervals or less. Figure 4-1 compares hourly water content changes measured using TDR, lysimeter, and Bowen ratio techniques over a 12-h period. The TDR-measured soil-water losses agree excellently with those of the other two techniques until 11:00, after which the results diverged. At the time of measurement, wheat roots had penetrated below 0.8

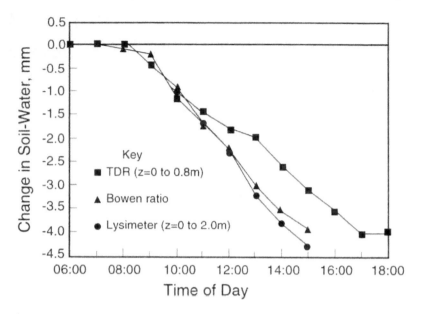

Figure 4-1. Hourly depletions of soil-water store measured using TDR, a weighing lysimeter, and the Bowen ratio technique under a wheat crop at Wagga Wagga, New South Wales (from Zegelin et al., 1992).

m, the depth of the lowest TDR probe. The TDR may have underestimated water losses because of this. A feasible interpretation of Figure 4-1 is that the wheat extracted water from depths above 0.8 m while evaporative demand was low before 11:00 and used deeper roots to extract water after this time (Topp et al., 1994).

NEUTRON PROBE

Schofield et al. (1994) used two field plots (3 x 10m) that had been monitored by neutron probe for more than five years for a comparison study with TDR measurement of soil water content. The plots were constructed of crushed Bandelier tuff and were uniform through the soil column to a depth well below that which was investigated. Each plot had three access tubes evenly spaced on the longitudinal axis of the plot.

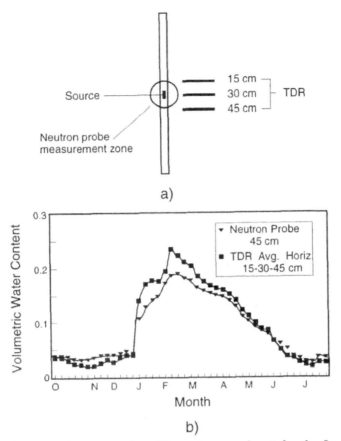

Figure 4-2. Comparison #2, neutron probe at depth of 30 cm: (a) cross section view; (b) time series plot (from Schofield et al., 1994).

Near two of the tubes on each plot, a series of TDR waveguides and temperature probes were placed at various depths with either a vertical or horizontal orientation to approximate the neutron probe interrogation zones. A 70 cm deep trench was excavated near each of the four access tubes. In the vertical face of each trench, two-rod TDR probes (30 cm long) were driven horizontally at depths of 5, 15, 30, 45, and 60 cm. Three are shown in Figure 4-2a. On the surface near each trench, two-rod TDR probes 15, 30, 45, and 60 cm long were driven vertically into place. The rod pairs were connected to multiplexers linked to a cable tester. The cable tester and multiplexers were controlled by a datalogger which also collected data. Temperature probes were placed in the access trenches at depths of 15, 30, 45, and 60 cm and connected to the same datalogger. The access trenches were backfilled with the original material and the surface of the plots restored to their initial condition as much as practicable.

A Campbell Pacific Nuclear Corp. model 503DR hydroprobe was used to collect neutron probe data. The probe uses an encapsulated Am^{241}/Be source and the calibration employed for converting the gross counts to volume percent water content was derived from previous monitoring of these test plots. Measurements were made at depths of 15, 30, 45, and 60 cm in the four access tubes near the TDR instrumentation, as well as in two tubes that were left undisturbed for control purposes. Measurements were performed within one hour of the TDR measurements on the same weekly schedule.

For purposes of comparison, it was assumed that the neutron probe measured a spherical volume of approximately 300 mm diameter (Gardner, 1986; Baker and Luscano, 1989), while TDR interrogated a cylindrical volume along the length of the waveguides. Also, it was assumed that moisture content distribution is laterally constant over the small footprint of the instrument array around each access tube ($<1m^2$). These assumptions allow comparisons to be made between the two measurement techniques since they sample very different volumes.

Two sets of comparisons were made at three depths in the soil column for a total of six comparisons at each of the four instrumented tube locations. Results for a particular comparison from all four locations were analyzed as a composite. Comparison #2 consisted of the neutron probe at 30 cm depth compared to the average of horizonal TDR values at 15, 30, and 45 cm depths as shown in Figure 4-2b. The computed regression equation between water content measured with the neutron probe, θ_{NP}, and that measured with TDR, θ_{TDR}, was

$$\theta_{TDR} = 1.231\theta_{NP} - 0.013 \qquad r^2 = 0.960 \qquad (4\text{-}1)$$

where r^2 is the correlation coefficient. However, readers must keep in mind that both θ_{TDR} and θ_{NP} are indirect measurements of water content with inherent uncertainties.

Previously, Dasberg and Dalton (1985) had conducted a field comparison of TDR and neutron probe measurements at the U. S. Salinity Laboratory. The test plots had dimensions of 3 x 3 x 1.5 m, included one neutron probe access tube, and the depth profile was accessible by plywood panels. Horizontal holes, 6.35 cm in diameter and 60 cm long were made at depths of 15, 25, 45, 75, and 105 cm through the side panels. Two-rod TDR probes (rod diameter of 3.2 mm and spacing of 50 mm) 200 mm or 300 mm long were carefully placed into the holes, which were then back-filled with the original soil. Gravimetric water contents, θ_m, were obtained from samples taken during installation of the TDR probes and when additional neutron access tubes were installed.

The calibration regression equations and correlation coefficients were

$$0_{TDR} = 1.02\theta_m - 0.023 \qquad r^2 = 0.84 \qquad (4\text{-}2)$$

$$\theta_{NP} = 0.86\theta_m + 0.031 \qquad r^2 = 0.82 \qquad (4\text{-}3)$$

$$\theta_{TDR} \; versus \; \theta_{NP} \qquad r^2 = 0.79. \qquad (4\text{-}4)$$

Dasberg and Dalton concluded that the correlation in Equation 4-2 is similar to that in Equation 4-3, but the former is somewhat closer to a 1:1 relationship.

Schofield et al. (1994) concluded that TDR appears to be a reasonable substitute for neutron probe measurements for the soil type and water content range investigated in their study with the exception of measurements made when the soil was frozen. The divergence of the two techniques when the water content was 20 percent and greater requires further investigation under laboratory conditions in which water content samples are obtained for comparison.

NUCLEAR DENSITY MOISTURE GAUGE

Look et al. (1994a) describe a case study of road construction in which both TDR probes and nuclear density moisture gauges were installed to monitor soil water content in order to assess the effects of saturation and other control parameters. The TDR probes were positioned in the middle third of each test section and located in the crushed rock pavement over sound subgrades.

The TDR probes were placed horizontally and installed directly on top of the subbase layer. Probe rods 500 mm in length were used to give the desired resolution in such granular material. The rods were covered with paving material to provide nominal cover ahead of the paving equipment. The pavement itself was a 200-mm base placed over a crushed rock subbase in some areas and a cement-treated subbase in others, at varying degrees of saturation and densities. Heavy compaction equipment was used to compact the base layer in which the TDR probes were installed.

Ten CSIRO probes were installed together with ten Queensland Transport (QT) B-probes (Table 3.2). After construction of the pavement, it was found that only 4 of the 10 CSIRO probes survived while all 10 of the QT B-probes survived. The B-probe was clearly better suited to the construction process. Figure 4-3 compares the moisture content change measured at each section using the TDR probes and the as-constructed conditions measured using a nuclear density moisture gauge. Good agreement was obtained. After three years, all 10 of the QT B-probes were still operational, while only 2 of the 10 CSIRO probes were operational.

The sensitivity of the volumetric moisture content (θ_v) reading to the variation in the calibration constants was examined. In general, a variation in θ_v of less that 0.5% was calculated for the QT B-probe while a variation of less than 0.2% was calculated for the CSIRO probe. The θ_v values recorded at these sites ranged from 0.10 to 0.20 as shown in Figure 4-3.

The accuracy of volumetric moisture content measurement

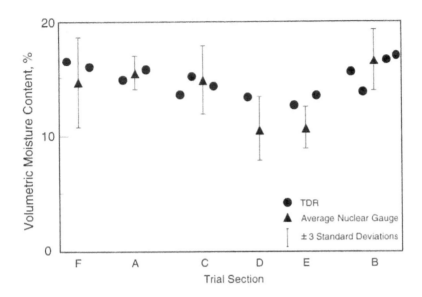

Figure 4-3. Comparison of initial readings for different probe designs (from Look et al., 1994a).

was considered to be acceptable for engineering purposes, where ease of monitoring is important. Change in moisture content was the important factor influencing design and construction rather than absolute moisture contents, so the data was considered acceptable. The confidence gained from the initial installations resulted in more than one hundred probes being installed in expansive clay embankments to monitor moisture changes responsible for movements and distress in pavement surfaces.

APPLICATIONS AND INSTALLATION CONSIDERATIONS

The potential for diverse applications of TDR monitoring of water content in soils and other porous media can be stimulated by considering a variety of ways in which it has been used in the past. Details of installation and interpretation which have been presented in the literature are an important experience database that should be utilized by current and potential users of TDR technology. Examples are presented for the following applications: agricultural irrigation control, landfill cover performance, infiltration studies, soil sampling, freeze-thaw behavior of soils, road and embankment construction on expansive soils, and pavement performance.

Agriculture–Irrigation Control

The TDR technique can be used to determine the "full point" (field capacity) water content of soil as described by Campbell and Campbell (1982) on a small plot chosen for that purpose (Topp and Davis, 1985). The determination of the full point water content is achieved by selecting a bare plot or plots about 1 m², having soil representative of the field to be irrigated, and installing three to five probes vertically within the plot(s) to the depth equal to the rooting limit. Each plot is flooded until there is no measurable increase in TDR-measured water content. Each plot is covered for 2 days to allow downward redistribution of water while evaporation loss through the surface is prevented. The resulting water content after 2 days is a good measure of the full value for the soil. The number of plots chosen for this determination is dependent upon factors such as soil variability, degree of precision required, and the labor available for the determination. In most cases, one well-chosen plot is sufficient.

It is also possible to set the full point in a field in which irrigation is already in use (Topp and Davis, 1985). In this case, the procedure to be described bears a lot of similarity to that given by Gear et al. (1977), who were using a neutron probe. The depth of soil over which the full point is to be determined is optional. The greater the depth chosen, the better it will represent the whole of the root zone. A depth of 0.3 m is practical. Three to five TDR probes reaching to this depth can be installed in soil and crop conditions representative of the field. Sufficient irrigation is applied to achieve or slightly exceed the full point of the chosen site to the depth of 0.3 m. This condition is achieved when the TDR-measured water content has ceased to increase during irrigation. Measurements of TDR water content values should be made at regular intervals, such as daily, after cessation of irrigation. As long as evaporative demand remains relatively constant, a plot of water content values against time will decrease in a regular fashion. After the first day, it is often possible to extrapolate the linear part of the drying curve back to the time of cessation of irrigation. The water content at that time on the extrapolated curve is the full point for the top 0.3 m of soil. The depth to which this full point applies may be extended by using a greater depth of irrigation and TDR probes of equivalent depth. Alternatively, one can assume, for uniform soil, that the top 0.3 m is representative of the whole rooting depth.

Setting the "refill" point, as described by Gear et al. (1977) and Campbell and Campbell (1982), is possible using a TDR hand probe (Topp et al., 1984). A portion of the water content vs. water potential relationship can be determined by simultaneous measurement of the water content by TDR and the water potential by tensiometer. In this instance, three tensiometers should be installed at two depths within the rooting zone, e.g., 75 and 175 mm deep. These two depths were suggested and chosen to represent zones within the cultivated soil and immediately below the depth of cultivation, if applicable. During drying of the soil within the tensiometer range, the water content is measured periodically over the depth increments of 50–100 mm and 150–200 mm by insertion of a TDR probe (Topp et al., 1984). The authors suggest that a 200-mm probe should be inserted in 50-mm increments and the water content determined after each additional increment. Progressive insertion as described by Topp et al. (1984) reduces the influence of spatial variability on incremental measurements. It is likely, however, that three measurements made near each tensiometer will assure greater reliability of the desorption data. Paired determinations of water content and potential should be taken when the tensiometers read between -10 and -20 cbar (centibars) and again between -30 and -50 cbar. The resulting data are plotted on three-cycle log-log graph paper, using separate sheets for each depth, and fitted by a straight line. Each line represents the desorption relationship for the soil at the depth of measurement. If the lines are similar, a single line may be used for this soil. If, however, the two layers have measurably different desorption relationships, it may be necessary to choose a separate refill point for each depth.

The desorption relationship can often be represented by an equation of the form

$$\psi = a\theta_v^{-b} \tag{4-5}$$

where ψ is the soil water potential (bars) and a and b are constants to be determined from the graphs. Campbell and Campbell (1982) have given a the value -4×10^{-4} bar and obtained b from field measurement using only one measured point.

The optimum soil water potential at which to commence irrigation can be chosen from published values to maintain maximum yield for particular crop conditions. Substitution of this potential into Equation 4-5 gives the limiting water content at which irrigation should commence. A separate value is found for each depth (75 or 175 mm). If these values are measurably different, it is advisable to find a linear weighted average value to reflect the relative depths of the two layers within the crop's root zone. In addition to the possibility of accounting for the effects of layering, it is often advisable to use a range of refill values for varying evaporative demands (Campbell and Campbell, 1982; Gear et al., 1977). Farmers with experience will be able to refine and adjust the refill value or range of values to reflect their particular soil and crop conditions.

Although vertical installation of TDR probes is appropriate for setting the full and refill points for a soil, there are significant advantages to installations at 45° off the vertical for lines intended for longer term monitoring and measurement. Probes inserted at an angle go across vertical inhomogeneities such as vertical drying cracks, worm channels, and rooting patterns of local extent. Objects placed vertically in soil tend to initiate drying cracks or holes which act as preferential paths for water during rainfall or irrigation (Topp and Davis, 1985). Probes installed at an angle have a reduced tendency to initiate cracks and openings. It is only slightly more difficult to install TDR probes at an angle using a guide for the rods than it is to install them vertically. Topp and

Davis (1985) used a 100-mm channel steel as a guide trough for installation of TDR probes consisting of a pair of 6-mm-diameter rods spaced 50 mm apart.

A number of TDR probes can be placed in a field and monitored continually or at preset times throughout the growing season. A control unit turns on the TDR pulser/sampler, measures each of the probes, and averages the readings. If the soil water content is at or below the refill point chosen by the farmer, then the irrigation system is switched on. The unit would make periodic measurements during irrigation until the water content reached the selected full value and the irrigation would be switched off.

Landfill Cover Performance

The Los Alamos National Laboratory examined water balance relationships for four different landfill cover designs containing hydraulic and capillary engineered barriers in 1 x 10 m plots with downhill slopes of 5%, 10%, 15%, and 25% (Nyhan et al., 1994). Seepage was evaluated as a function of slope length for each plot, as well as interflow, runoff, and precipitation, using an automated waterflow data logging system that routinely collected hourly data. Soil water content within sixteen field plots was monitored four times a day using TDR techniques with an automated and multi-plexed measurement system. Volumetric water content was measured with 600-mm long waveguides at each of 212 locations. One set of waveguides was emplaced vertically in four locations in every soil layer to determine soil water inventory in each field plot; a second set of waveguides was emplaced horizontally in several soil layers to provide a more detailed picture of soil water dynamics close to soil layer interfaces. TDR monitoring made it possible to detect pulses of soil water moving through the topsoil and engi-neered barriers with much higher temporal and spatial resolution than could be achieved with neutron moisture gauges. This meant that landfill cover design factors such as slope and slope length could be assessed in terms of the performance of the four designs used in the field study.

Benson et al. (1994) constructed and instrumented three test sections to assess the hydrologic behavior of earthen final covers at two municipal solid-waste landfills. A traditional design using a resistive barrier was used for two of the test sections, whereas an alternative design using a capillary barrier was used for the third test section. The test sections were built in two distinctly different climates: humid with high precipitation (Atlanta, Georgia) and arid (East Wenatchee, Washington). Each test section was instrumented to measure climatological variables, overland flow, percolation, soil temperature, and soil water content.

Soil water content was monitored using TDR techniques. Three nests of 300-mm long two-rod probes were placed along the centerline of each test section; they were located at 5, 10, and 15 m upslope from the bottom berm. In Atlanta, four probes were placed in each nest (depths = 0.15, 0.38, 0.61, and 0.84 m), whereas five probes were placed in each nest at Wenatchee (depths = 0.08, 0.23, 0.38, 0.53, and 0.68 m). Four measurements of water content were conducted every minute; a full set of measurements was typically made once per hour.

A data acquisition and control computer was used to collect data and control various components of the monitoring system. The system in Atlanta was powered by 115 VAC whereas the system in Wenatchee was powered by solar panels and batteries. Telecommunications were established via traditional telephone lines or by cellular transmission. Commercially available equipment sold by a variety of vendors was used.

The City of Glendale, Arizona is using TDR systems to evaluate two final cover designs (Boehm and Scherbert, 1997). The currently approved monolithic design consists of 60 cm of expensive imported clayey soil overlain by 15 cm of vegetative support soil and the capillary break design consists of a geomembrane with an overlying layer of protective soil. Less expensive alternative covers are being evaluated using two 60-m by 60-m test plots constructed using locally available silty sand soils. One test plot was constructed without vegetative cover, thereby simulating the sparse vegetation of a desert environment. The second plot was constructed with nonnative grass cover and a water irrigation system. One-half of each test plot (30-m by 60-m) was constructed with a monolithic cover design and the other half was constructed with a capillary break cover design.

Each test plot has a dedicated soil moisture monitoring system and multiplexer network. The monitoring system employs TDR probes manufactured by ESI Environmental Sensors, Inc. which consist of two 1.5 m stainless steel bars divided into segments to monitor multiple depths (Figure 10-3). The probes have an active length of 1.2 m with five discrete measured intervals or segments. Shallow probes extend from the surface to 1.5 m and monitor moisture at the following depths: 0–15 cm, 15–30 cm, 30–60 cm, 60–90 cm, and 90–120 cm. Deep probes are buried approximately 60 cm and monitor moisture at depths of 60–75 cm, 75–90 cm, 90–120 cm, 120–150 cm, and 150–180 cm. The probes are situated in six clusters in each test plot.

The system is currently programmed to read the soil moisture probes once per hour. The internal data storage unit is downloaded approximately weekly onto diskettes via a portable computer. A copy of the weekly data file is copied for on-site storage prior to sending it to project personnel for inclusion into the project database. The field data collection effort will indicate a limiting depth for significant/measurable moisture fluctuations and thus provide real evidence that percolation through the final cover is extremely low. Computer modeling is used to calculate the actual percolation and thus provide a means for extrapolating soil moisture behavior over a 30-year post-closure period.

The soil moisture data is being reviewed for compatibility with fundamental laws of soil physics. At a minimum, this includes validation that: a) soil moisture within the upper six inches increases during and/or immediately after rainfall events, b) soil moisture decreases during prolonged hot dry periods, and c) measured volumetric moisture contents ($\theta_v = V_w / V_{tot}$) are not typically greater than the porosity ($\eta = V_v / V_{tot}$) of the soil. In general, the field data reviewed thus far are consistent with the above criteria. However, there have been several exceptions, which can be attributed to individual probe performance. These particular probes are continually being monitored or have been replaced, and calibrations between volumetric moisture content and TDR probe values are being refined (Benson, 1997).

Infiltration and Wetting Fronts

Topp et al. (1982) performed an extensive set of laboratory studies
to assess the capability of TDR to monitor water contents in cases
where there were steep water gradients and progression of wetting
fronts.   Air-dried soil was packed around vertical probes in
cylinders and water was added to the top. By virtue of reflections
which developed at the wetting front and at the end of vertical
probes, it was possible to track progression of the front. Topp et
al. (1983) describe field studies at two sites in which probes 0.15,
0.30, 0.50, 0.80, and 1.20 m long were installed vertically. Also,
pits were dug and 0.50-m long probes were installed into the pit
walls at depths of 0.075, 0.225, 0.40, 0.65, and 1.00 m. Comparisons between recorded rainfall and measured soil water content
showed the possibility of using TDR probes to measure the
infiltration of natural rainfall and detect the progression of the
wetting front.
        Knowlton et al. (1994) expanded on these earlier studies to
define the wetting front advance. The following approach was
investigated. The lengths of the wet zone (1_1), transition zone
(Δz), and dry zone (1_2) shown in the upper part of Figure 4-4 were
calculated using two TDR probes (long probe, 1_3, and short probe,
1_4), the apparent length before infiltrating (l_{a3}), and the apparent
length after saturation (l_{a4}). The transition zone was calculated as

$$\Delta z = 1_3 - 1_2 - 1_1 \tag{4-6}$$

based on TDR measurement of $1_1 [= 1_{a1} * (1_4 / 1_{a4})]$ and $1_2 [= 1_{a2} *
(1_3 / 1_{a3})]$. This transition zone was used as input to a computer
model, along with other information to determine hydrologic
parameters.

Soil Sampling

Kaya et al. (1994) proposed that water content of soils could be
determined in a borehole by employing a modified split-spoon

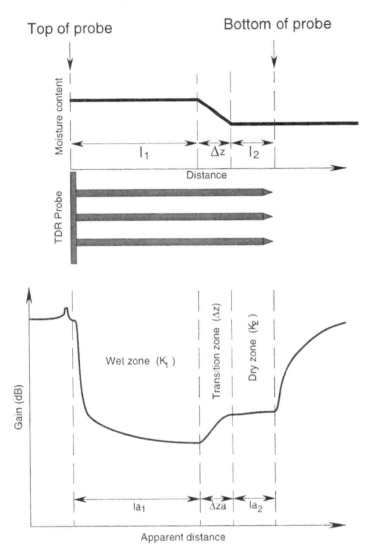

Figure 4-4. Wetting front advance application: (a) idealized moisture content distribution during infiltration, and (b) the TDR trace used to quantify the distribution of moisture (from Knowlton et al., 1994).

sampler with a rod in the middle to create a coaxial cell. When the split-spoon sampler is in the borehole, it can be connected to a TDR system and in-place water content determined. As a practical matter, this design is not feasible, but a variation of the central rod could be implemented. Knowlton et al. (1994) described the use of a miniaturized TDR probe to measure water content of samples before, or immediately after, being extracted from the soil sampling tube. Knowlton (1994) described work being done at Sandia National Laboratories to integrate a TDR probe into a cone penetrometer to allow determination of *in situ* water content in soils.

Road and Embankment Construction on Expansive Clay

Knowledge of in-service moisture variations is especially important for roads constructed on volumetrically active (expansive) clays (Look et al., 1994b). All clays shrink or swell when subjected to moisture variations. Those that move to a considerable degree and/or affect the performance of a structure (building, road, culvert, etc.) are termed "volumetrically active" or "expansive." Pavements constructed on expansive clays experience two main types of primary distress: 1) pavement undulations (distortion) and 2) longitudinal edge cracks. When these distresses are combined with traffic loading and/or further climatic influences, such as ponding of water or water entering the cracks, the pavement may then experience structural distress such as rutting and the formation of potholes.

TDR has proven to be a useful tool to define certain expansive clay characteristics. In an expansive clay soil, both the porosity ($\eta = V_v / V_{tot}$) and degreee of saturation ($S = V_w / V_v$) may change as the clay shrinks and swells. While porosity and saturation are interdependent, it is often useful to examine these parameters independently. Measuring volumetric water content ($\theta_v = V_w/V_{tot}$) with TDR in a saturated clay ($S = 100\%$) provides information on the volume change (porosity changes). In an unsaturated clay with fully constrained conditions ($\eta = $ constant), changes in θ_v provide some indication of saturation changes. The actual change in θ_v which can be directly attributed to porosity or degree of saturation would lie between the two extreme conditions. The behavior of expansive clays is also associated with soil suction (moisture potential) changes, which can be correlated with changes in θ_v.

Expressing moisture content on a volume basis can therefore provide information on 1) porosity changes, 2) degree of saturation, and 3) soil suction behavior.

Embankments were to be constructed from expansive clay material excavated from adjacent cuts (Look et al., 1994b). In order to determine the in-service conditions to be expected along the route, an investigation was undertaken on existing road embankments. The shoulder pavement was removed to expose the embankment subgrade and two trenches were excavated. Dynamic cone penetrometer tests were performed to evaluate the existing California Bearing Ratio (CBR) conditions with depth. *In situ* density tests were performed, and tube samples were obtained to determine soil matric suction (by the filter paper method), density, moisture content, and shrink/swell characteristics. Bulk samples were also taken to determine the density ratio for standard Proctor compaction. TDR probes were installed at various levels in the trench walls to provide data on existing and long-term moisture conditions and establish the approximate active and stable zones.

Good agreement was obtained between the moisture content obtained using TDR and measurements of moisture content by gravimetric means (Look et al., 1994b). The results of TDR monitoring are compared with rainfall in Figure 4-5 for the stable zone. The in-service moisture content was at about 1.2 times the optimum moisture content (OMC) in the dry season condition and at 1.4 times OMC in the wet condition. The approximate boundary between the active and stable zones was determined from results of TDR monitoring as well as oedometer testing aimed at establishing the swelling characteristics of samples at various levels. The zero swelling strains obtained for samples taken in trenches at depths of 1.85-m and 1.80-m indicated that at this depth the soil was approximately in an equilibrium condition. The stable zone is determined when the existing overburden pressure ≥ zero swell pressure. The moisture content range of 1.2 to 1.4 OMC and density range of 0.92 to 0.94 maximum dry density (MDD) were then used to obtain laboratory compaction data to determine the design CBR.

Three trial embankment sections were constructed, each at different moisture conditions but with the same target density. The

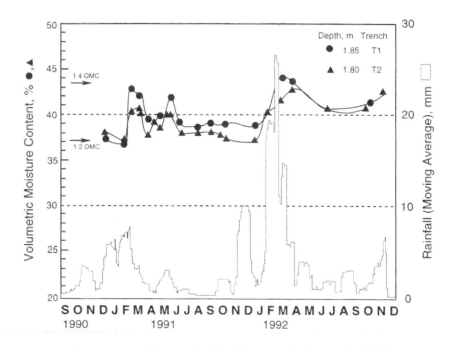

Figure 4-5. Cooroy trench monitoring–stable zone. Rainfall moving average is for 10 days before and 10 days after event (from Look et al., 1994b).

sections constructed were 1.0 OMC, 1.2 OMC, and as wet as possible (1.2+ OMC). TDR moisture probes were installed at three levels within each trial embankment section. The probes were located alongside movement plates used to monitor movements within the embankment. The intent was to evaluate movement changes at both the inner wheel path (IWP) and the outer wheel path (OWP) positions at the different levels. Figure 4-6 shows moisture changes experienced by trial sections in the stable zone. The trend indicates that the drier section had swelled and that within 9 months of construction, the three embankment sections converged to above 1.2 OMC (the wet of optimum condition) coinciding with the wet period rainfall.

During actual construction, TDR and other instrumentation were installed to 1) monitor moisture changes and volumetric

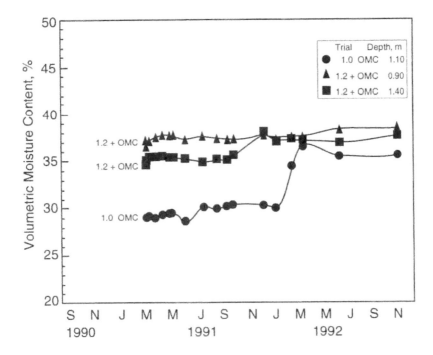

Figure 4-6. Trial embankments equilibration–stable zone in inner wheel path (from Look et al., 1994b).

movements over time, 2) provide data to construction managers regarding achievement of target conditions and so allow decisions to be made regarding recommencement of pavement construction, and 3) provide long-term (several years) data to allow refinement of shrink/swell prediction models for roadway structures. Instruments were installed at various levels in the embankment. A typical layout is shown in Figure 4-7. Horizontal inclinometers were used to monitor movements accompanying the moisture changes being measured by TDR probes. Thermocouples were also installed at selected sites to measure the effect of temperature variation on moisture changes. Periodic road roughness measurements and survey leveling were used to evaluate the performace of the overlying pavement.

Freeze–Thaw Behavior of Soils

TDR has proven to be a useful tool in the field of frozen soil

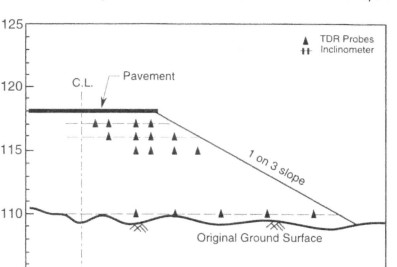

Figure 4-7. Typical installation layout. TDR probes to monitor the moisture changes and horizontal inclinometers to monitor the associated vertical movements within the embankments (from Look et al., 1994b).

hydrology, because the method is nondestructive and allows measurements with high time resolution (van der Keur, 1995). Still, some problems may occur when using TDR in a soil subjected to freeze–thaw cycles for freezing of a soil causes the soil to crack similar to drying of a soil. This may lead to air pockets between the TDR probe and the soil with erroneous measurements as a result. This is particularly a problem when studying infiltration of water, i.e., snow melt, into a partially frozen soil.

Patterson and Smith (1980) conducted a series of experiments to determine volumetric unfrozen water content with TDR-measured dielectric properties. Soil samples were placed in a thin-walled PVC tube 20 cm long (4.5 cm ID) covered with a latex membrane. This tube could be immersed in a circulating methanol bath which permits temperature control for extended periods of time. A two-prong probe of 0.32 mm diameter stainless steel rods with a length of 175 mm and spacing of 25 mm was used. The saturated samples were ramped through a temperature cycle and K_a was determined at each temperature. A curve of K_a versus temperature took about 4 days to compile for a heavy clay and 2 to 3 days for a silty loam. On average, each point takes about 4 to 5 hours. However, in clay at temperatures between $0°$ to $-1°C$, where substantial phase change takes place, up to 24 hours are necessary to attain equilibrium in the sample. K_a can be monitored a number of times at each temperature until a steady state is indicated (i.e., no further change in the TDR trace). Since the dielectric constant for ice ($K_{ice} = 3.2$) is very similar to that of dry soil, a soil containing ice closely resembles dry soil in a dielectric sense. For example, K_a for air-dried Ottawa sand was determined to be 2.9. When it was solidly frozen ($-18°C$) with $\theta_v = 0.24$, K_a was 3.4. It is suggested, therefore, that the observed variation in K_a with negative temperature is actually a response to the accompanying variation in unfrozen water content. Since this could not be directly confirmed, an experiment was devised to measure the liquid water contents in various ice/water mixtures.

The experiment entailed adding known volumes of previously supercooled water to a coaxial tube of known volume containing finely crushed ice. For the various volumes of water added, K_a was determined. The time required to obtain a reading after a volume of water was added was less than 30 seconds. The water was drained after each reading to check the volume. The curve of K_a versus θ_{uf} for the ice/water mixtures fell within the 95% band of the K_a versus θ_v curve determined by Topp et al. (1980) for a variety of frozen soils.

Baker et al. (1982) compared the TDR technique with the temperature measurement method for locating the frozen–unfrozen interface in water and sandy soils. This technique depends upon the high frequency (1 MHz to 1 GHz) electrical properties of water that change significantly and abruptly between the liquid and solid phases. Two-rod TDR probes were inserted into the soil. The frozen–unfrozen interface produced reflections measured by the TDR which were in turn used to locate the interface as it moved along the probes. In the laboratory it was possible to locate the interface to within ±0.5 cm and in the field to within ±2.4 cm. These errors were equal to those associated with the temperature measurements.

Stein and Kane (1983) used TDR to monitor the unfrozen

water content in soil. They also utilized this technique to determine snowmelt infiltration into seasonally frozen soils, and explored the feasibility of using TDR to monitor snowmelt percolation in the snowpack. Various configurations of parallel rod probes were installed horizontally at various depths in the soil and also in the snowpack. This technique gave good delineation of temporal changes in the profile of unfrozen water content versus depth. Results looked promising in snow if *in situ* snow density measurements were taken along with the TDR measurements, which is consistent with the discussion of soil density influence in Chapter 3.

Baker and Allmaras (1990) used a multiplexed system with 12 probes to observe infiltration at multiple points at a field site. This allowed measurement of unfrozen water content as a function of space and time during freezing and thawing of the soil profile. The 300-mm-long stainless steel probes were installed horizontally into the wall of a pit at depths of 100, 300, and 500 mm with four probes in a row at each depth. The system has also been used to estimate the reproducibility of water content measurement by TDR, which was found to be in the range of ± 0.006 to ± 0.008 m^3/m^3

Freeze–Thaw Pavement Performance

The capability of monitoring freeze–thaw behavior makes TDR applicable for assessing factors that influence pavement performance. The influence of temperature and moisture condition on pavement performance can readily be seen in seasonal frost areas, which cover about one-third of the continental United States, nearly all of Alaska and Canada, and substantial portions of Northern and Eastern Europe and northern Asia (Janoo et al., 1994). During the winter, heat is lost from the pavement structure and subfreezing temperatures draw water through frost-susceptible soil to the freezing front. At the freezing front, the water forms ice lenses. The bearing capacity of the frozen pavement structure is increased. This annual occurrence can be used by smaller airports to land heavier planes during the winter. On the damaging side, the ice lenses and variant subgrade can produce differential frost heave. This can be seen on some highways, but it is more common on secondary roads.

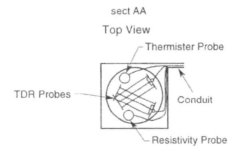

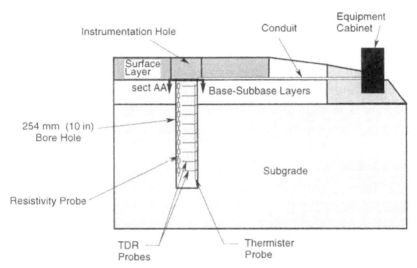

Figure 4-8. Illustration of instrumentation installation (from Rada et al., 1994).

During winter and spring thaws, the ice lenses melt, producing excess moisture in the pavement structure, which results in a dramatically reduced bearing capacity of the structure. These periods of thaw-weakening depend upon the soil type, degree of saturation, and drainage conditions. Some pavement structures designed for 80-kN axle loads are unable to carry the load during the thaw period. During the spring thaw, load restrictions are commonly placed on these roads and the duration of the load restriction is based on judgement and experience. The U.S. Army Cold Regions Research and Engineering Laboratory (CRREL) has developed a theoretical coupled heat and moisture model for predicting frost heave in a given pavement structure. The model (Guymon et al., 1993) also predicts temperatures and the movement of water in the pavement structure. Moisture measurement using TDR was introduced to pavement engineering around 1989 (Neiber and Baker, 1989; Neiber et al., 1991). CRREL has installed TDR systems throughout the country using pits excavated beneath the pavement shoulder and using variations of the auger hole installation shown in Figure 4-8. There have been some problems. At a site in Vermont, the volummetric water content measurements in the base course were inconsistent because, in many cases, they were higher than the soil's void volume. Janoo et al. (1994) concluded that this was a problem with the algorithm used to calculate travel time based on tangents to the TDR waveform which emphasizes the necessity of site-specific calibration.

An endeavor referred to as the Seasonal Monitoring Program has been undertaken as part of the Federal Highway Administration's Long-Term Pavement Performance studies (Rada et al., 1994). The objective of this program is to provide the data needed to attain an understanding of the magnitude and impact of temporal variations in pavement response due to the effects of temperature and moisture variations and traffic. To gather much of these data, the program is relying on instrumentation installed at 64 pavement test sections throughout North America. At each site, TDR probes were installed as shown in Figure 4-8 to monitor moisture changes. A three-prong probe was refined and fabricated by FHWA for use in the program.

Kotdawala et al. (1994) made TDR soil moisture measurements at sites on Kansas State Route K-18 and U.S. Interstate I-70. TDR was found to be more convenient for pavements as the probes could be buried under the existing pavements and intermittent readings of soil moisture could be taken without disturbing traffic. At each location, 3-prong probes were installed under the pavements. A 100-mm diameter core hole was drilled on the outer wheel path of the travel lane extending to the subgrade and the probe was driven into the subgrade using a custom-made installation head. After driving, the top of the probes was approximately 460 mm beneath the pavement surface. On concrete pavements, drilling was stopped before the drill bit reached the underlying layer to prevent intrusion of drilling water into the subgrade and the access hole was completed using a hammer. For asphalt concrete pavements, access holes were drilled without water. An 8-m coaxial cable was extended from the probes to the edge of the shoulder along saw-cut grooves in the paved surfaces. The saw cuts were sealed with silicone-type joint sealants. The end of each cable was secured in a housing which could be accessed very easily for moisture measurement. After installing the waveguides, excavated soil was replaced and compacted.

The pavement cores were replaced and core holes sealed to prevent any water intrusion. Despite frequent precipitation, it was found that the soil-moisture contents did not change significantly at either site. The relatively constant moisture condition under paved surfaces was expected since the surfaces are virtually impervious, and soil-moisture movement is negligible during the relatively dry winter months in Kansas. One sensor on I-70 registered unusually high moisture content in March, possibly due to the combined effect of snow melt and perched water at that location.

Chapter 5
MONITORING LOCALIZED DEFORMATION IN ROCK

Measuring rock deformation with TDR developed naturally from the original purpose of this technology—identification of cable faults or deformities along telephone and power cables. TDR cables were first grouted in rock masses around mining operations simply to locate cable breaks which correlated with the extent of complete rock mass failure. During one such application it was noticed that the TDR reflection waveforms changed incrementally with rock mass deformation before failure. Subsequent laboratory studies indicated that it was possible not only to quantify the magnitude of deformation but also, in some cases, to distinguish shearing from tensile deformation. Since that time, cables have been installed to monitor rock mass movement in a variety of situations. This chapter presents variations in waveform characteristics which develop with cable deformation, cable calibration techniques, and installation techniques for metallic cables grouted in rock.

GENERAL INSTALLATION CONSIDERATIONS

Coaxial cables are grouted into boreholes in a rock mass as shown in Figure 5-1, where shearing along joints deforms the rock mass as well as the cable. Ultra-fast rise-time voltage pulses are sent down along the cable from a TDR cable tester. Cable defects such as shears, extensions, and ultimately complete severance produce reflections which are detected by the tester. Pulse reflections from all changes in geometry along the cable are superimposed on the input pulse to form a reflected TDR waveform. Consequently, the waveform consists of many individual reflections associated with localized deformations along the cable. Characteristics of a TDR

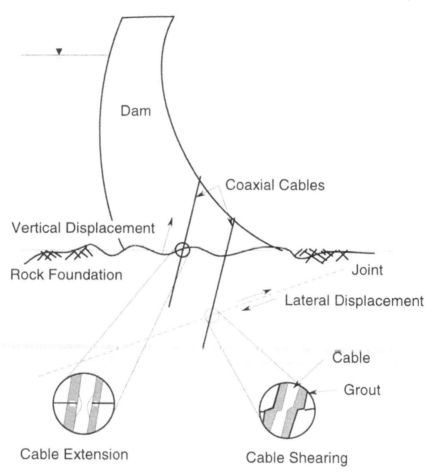

Figure 5-1. Extension and shearing of coaxial cable resulting from relative vertical and lateral dam displacements, respectively.

reflection are determined not only by the magnitude of cable deformation, but also by the type of cable defect.

Features of cable installation such as hole size, cable type, cable diameter, crimping, grout properties, etc. can be varied to meet the particular needs of a project. Prior to installation, the cable is crimped at several appropriate locations to optimize location accuracy and an initial reading is taken to ascertain that all crimps will be recorded. The cable is lowered down a hole and bonded to the surrounding rock with an expansive cement grout that is tremied into the hole.

Reference Crimps Improve Location Accuracy

Reference crimps improve the accuracy with which cable defects can be located. Without reference crimps along a cable, accuracy is on the order of 2% of the distance from the tester, so a defect at a distance of 30 m (100 ft) can be located to within 0.6 m (2 ft) of its actual position. If a cable is crimped at known locations, subsequent cable defects created by rock mass movement can then be located relative to these precisely known reference crimps. See the discussion of crimp width in connection with the discussion of attenuation later in this chapter.

INTERPRETATION OF REFLECTIONS

In order to interpret the characteristics of a TDR reflection in terms of magnitude and type of cable deformation, it is necessary to calibrate reflections for different modes of cable deformation. There are two major categories of relative movement between rock blocks: shear displacement and normal displacement. Shearing occurs when two adjacent blocks move parallel to their interface. Normal movement occurs when two blocks move perpendicular to their interface. These typical scenarios of shearing and extension are shown in Figure 5-1.

Shear of two blocks relative to one another distorts the cable cross section as shown in the insert in Figure 5-1. The change in cable geometry causes a local capacitive cable fault, the magnitude of which is inversely proportional to the distance between the outer and inner conductors (Equation 2-21). By determining the characteristic of a single shear displacement under controlled laboratory conditions, it is possible to extrapolate the results to the field in order to estimate relative shear movement of adjacent strata in which a coaxial cable has been grouted.

When adjacent blocks move perpendicular to their interface as shown in Figure 5-1, the cable undergoes extension and contracts in diameter as it necks down. The local capacitance of the cable at that point will increase if the decrease in diameter of the outer conductor is greater than the decrease in diameter of the inner

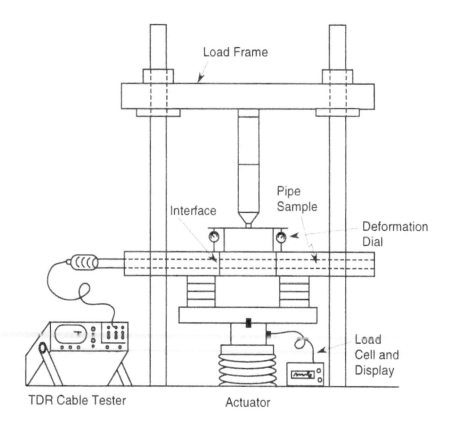

Figure 5-2. Set-up for cable shear test (from Su, 1987).

conductor. For the case in which there is combined shear and extension of a cable, the effect will be additive and create a more complex condition.

Simple Shearing

The type and magnitude of TDR reflection caused by shear and extension have been investigated in the laboratory by several researchers (Table 5.1). For shearing, a bounding pipe and grout (but not the cable) were cut in two places to produce three segments as shown in Figure 5-2. Each segment represented a rock block while the transverse cuts simulated joint planes. The grout and embedded cable were sheared by pushing the middle segment downward while restraining the two end segments. During the test, relative movement between pipe segments and the associated TDR waveform were recorded either with a chart recorder or in a digitized form. The force-deformation behavior during a shear test conducted by Peterson (1993) using 12.7-mm-diameter cable is shown in Figure 5-3. The initial force-deformation behavior is relatively stiff until failure of the grout occurs at the cuts. During an earlier series of tests, Su (1987) found irregularity of the subsequent load-deformation behavior was the result of the complex interaction between the grout and coaxial cable within the two shear zones as moments developed during the test. He found shear failure or breakage of the cable's outer conductor occurred at deformations of 80% to 90% of the cable diameter. The cable continued to resist load after failure of the outer conductor because the polyethylene dielectric and inner solid aluminum conductor remained intact. The peak force in Figure 5-3 is discussed later in this section and in the subsequent discussion of shear zone width.

A typical change in TDR reflection characteristics during shear is shown in Figure 5-4. It can be seen that reflection spikes form and then increase as shearing increases. Eventual failure of the outer conductor forms an open circuit and resultant positive reflection following the negative spike. As indicated by the test data plotted in Figure 5-3, the magnitude of the negative reflection spike increases with shear deformation but not until an initial deformation has taken place. This initial deformation necessary to fracture the grout and generate a detectable reflection depends upon the cable diameter and the distance from the cable tester to the shearing location.

Data from the tests similar to those in Figure 5-3 are plotted in Figure 5-5 as TDR reflection magnitude, V_{pk}, in mρ (see Figure 2-4) versus measured transverse displacement, u, in mm (which is consistent with notation used in rock mechanics). The correlation is assumed to be linear,

$$V_{pk} = Su + V_{pko},\qquad(5\text{-}1)$$

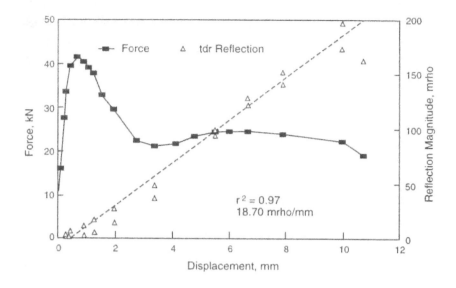

Figure 5-3. Applied force magnitude and TDR reflection magnitude versus shear displacement (from Peterson, 1993).

which is consistent with the relationship between change in capacitance and cable deformation shown in Figure 2-6. The slope, S, in units of mρ/mm is the sensitivity of the TDR reflection coefficient to cable deformation, and V_{pk0} in units of mρ is the y-axis intercept. The x-axis intercept, u_0, in units of mm is the displacement that must occur to fracture the grout and initiate cable deformation significant enough to produce a reflection,

$$u_0 = -V_{pk0} / S. \qquad (5\text{-}2)$$

So the TDR reflection magnitude can be converted to displacement by

$$u = (V_{pk} - V_{pk0}) / S = (V_{pk}/S) + u_0. \qquad (5\text{-}3)$$

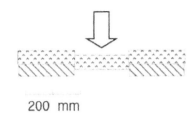

200 mm

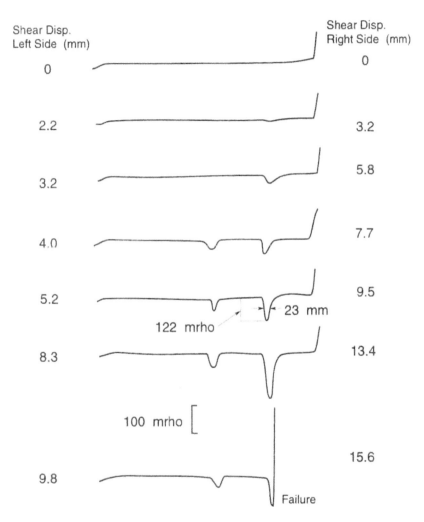

Figure 5-4. Increasing TDR reflection magnitude with increasing shear displacement (from Su, 1987).

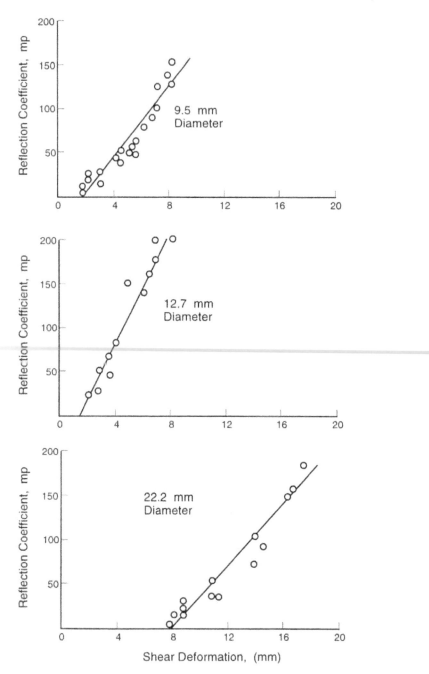

Figure 5-5. Correlation between TDR reflection magnitude, V_{pk}, and shear displacement, u, for short cables of different diameter (from Su, 1987).

This linear relationship is attractive due to its simplicity, but it is not necessarily the best model, and nonlinear correlations (Equations 6-1 and 6-2) have also been developed. This linear model has been used extensively to evaluate the sensitivity and reliability of TDR for quantifying the magnitude of deformation due to localized shear.

The sensitivity, S, of the linear best-fit to the reflection coefficient versus shearing deformation data in Figure 5-5 is 19.5, 31.0, and 16.5 mρ/mm for the 9.5, 12.7, and 22.2 mm diameter cables when shear occurred at a distance of 1.9 m from the cable tester. The initial deformation, u_0, was 1.58, 1.37, and 7.76 mm. While the two smaller cables were more sensitive to shearing, the larger cable withstood greater shear deformation before failure. Consequently, when choosing the coaxial cable appropriate for a project, a trade-off may be necessary among a) sensitivity to deformation, b) initial displacement required to produce a detectable reflection, and c) maximum displacement before shear failure.

Peterson (1993) conducted a comprehensive test program in order to independently evaluate the study done previously by Su (1987). It is important to note differences in the test details. For one series of tests, 12.7 mm diameter cables grouted into 75 mm diameter PVC pipe were subjected to shear. Twelve assemblies were prepared using a grout mix with 7.1% bentonite and twelve were prepared using grout without bentonite. After the grout had cured for periods of 5 to 17 days, the pipe was scribed around its circumference. This scribing left a pipe wall thickness of 1.0 mm (0.040 inch). By comparison, Su (1987) had cut completely through 51 mm diameter steel pipe to a depth of 2.5 mm (0.1 inch) into the grout, so the peak force (6.7 kN) was smaller than that shown in Figure 5-3.

The data points for all of Peterson's twenty-four grouted cable tests are presented in Figure 5-6, where the slope of the best-fit linear correlation is 19.7 mρ/mm (versus Su's value of 31.0 mρ/mm in Figure 5-5). Scatter and correlation reliability are as important as the sensitivity. The dashed lines represent envelopes for the 95% and 75% confidence intervals. The reliability for these particular laboratory conditions is ±2 mm with a 75% level of confidence, but only ±5 mm with a 95% level of confidence.

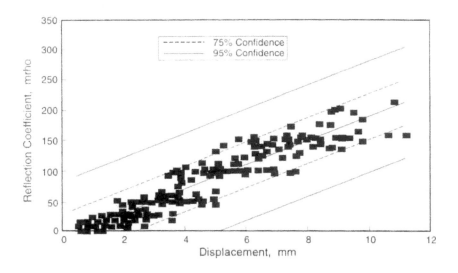

Figure 5-6. TDR reflection data from a series of laboratory direct shear tests (from O'Connor, 1991).

Effect of Shear Zone Width

To simulate a variable shear zone width, Peterson (1993) placed 89 mm (3-1/2 inch) diameter grouted cable assemblies in a special clamping device that allowed a wide variation of gaps between bounding clamp sections. The gap between the central clamp and left clamp was changed for each test. Six tests were performed using 19.1 mm diameter CommScope cable in which the gap width was 5 mm, 20 mm, 40 mm, or 80 mm. Grouted cable assemblies had been cured over periods which varied from 5 to 71 days.

Although the correlation between TDR reflection magnitude is not necessarily linear, a summary of best-fit linear correlations for the range of shear zone widths is shown in Figure 5-7, Table 5.1a, and Table 5.1b. The sensitivity (i.e., slope of line) decreased from 14.9 mρ/mm to 1.0 mρ/mm as the gap width increased from 5 mm to 80 mm. Such a decrease in magnitude of sensitivity is similar to

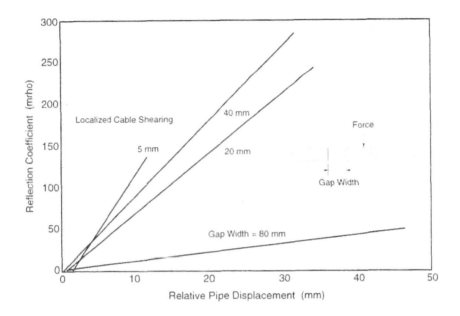

Figure 5-7. Influence of shear zone width on TDR reflection magnitude (from O'Connor, 1991).

that experienced with field installations (Peterson, 1993; O'Connor, 1991).

TDR signatures such as those in Figure 5-4 will change significantly as the gap width is increased. The mode of grout fracturing, cable deformation, and cable failure varies from localized shearing to bending of cable through a shear zone. As the gap width increases, the TDR reflection not only increases in magnitude but also in width (i.e., wavelength). In Figure 5-4, the spike magnitude is 122 mρ and spike width at half-amplitude is 23 mm at a shear displacement of 9.5 mm. For a gap width of 80 mm, at the same displacement, the magnitude is only 27 mρ and width at half-amplitude is 36.6 mm. The greater width-to-magnitude ratio is due to the difference in cable deformation: bending with larger gap widths versus localized shear with smaller gaps. When the gap was reduced to less than 5 mm, localized shearing of the cable was evident. Effects of bending are discussed further in Chapter 7.

It is important to pay attention to magnitudes of force and displacement in Figure 5-3. Assuming that the two reaction forces acting upward in Figure 5-2 are equal, their magnitude is one-half the applied downward force. So, a force of 1/2 (41740 N) [= 5218/2 lb] was required to initiate fracture of the grout surrounding a cable in a 76 mm diameter PVC pipe. This translates into an average shear stress of 4600 kPa [= $(41740/2 \text{ N})/(\pi/4D^2)$ = 665 psi]. By comparison, for the test with a shear zone width of 40 mm, a force of only 9875 N (2470 lb) was required to initiate fracture of an 89 mm diameter assembly, which translates into an average stress of only 1590 kPa (230 psi). In tests performed by Su (1987) with 51 mm diameter steel pipe molds which were completely scribed, fracturing was initiated at a force of only 1480 N (370 lb), which translates into an average shear stress of 1010 kPa (146 psi).

The initial high peak stress is partially a reflection of the PVC pipe not being cut completely but only scribed, as discussed above. However, the residual stress magnitude of 2000 to 2500 kPa is consistent for both Peterson (1993) and Su (1987), which indicates that residual resistance is associated with the cable without any influence of the shear frame or bounding pipe mold. There was not a significant difference in the average peak stress with 7.1% and 0% bentonite (5303 kPa and 5102 kPa) or the average residual stress (1907 kPa and 2115 kPa).

Tests on bare cable were performed using 12.7 mm Cablewave cable (O'Connor, 1991; Peterson 1993) to model debonding in the movement zone whereupon the fractured grout does not have adequate confinement. A modified direct shear frame was fabricated which allowed for clamping of bare cables. The maximum force was approximately 3100 N, which is consistent with the magnitude of the residual force attained in grouted cable tests shown in Figure 5-3. This consistency indicates that the gradual decrease from a peak force to a residual force is associated with transfer of stress from the grout to the cable with a reduction of any frictional resistance provided by the fractured grout. Rather than experiencing localized shearing, the bare cables were bent across a gap between the shear device clamps.

Table 5.1a. Laboratory direct shear calibration of grouted cables.

Cable	Diameter (mm)	Linear Regression			Comment	Reference
		slope, S (mrho/mm)	x-intercept, u_0 (mm)	r^2		
A	9.5	19.5	1.8	0.8987		Su (1987)
		22.8	1.0	0.9383		Kim (1989)
B	12.7	31.0	1.2	0.9100		Su (1987)
		9.7	1.8	0.8282	fracture/debond	Peterson (1993)
		24.8	2.4	0.9944	no fracture/debond	
		19.7	0.5	0.9999		O'Connor (1991)
		20.0	0.5	0.9354	7.1% bentonite	Peterson (1993)
		19.5	0.4	0.8947	0% bentonitie	
		15.1	1.4	0.9568		Kim (1989)
C	22.2	16.5	7.9	0.9360		Sun (1987)
		18.8	6.1	0.9465		Kim (1989)
D	13.7	13.9	0	0.9640		Aimone-Martin et al. (1994)
E	12.3	25.9	0	0.9810		

Cables
A = Cablewave FXA38-50, smooth aluminum outer conductor, foam polyethylene dielectric
B = Cablewave FXA12-50, smooth aluminum outer conductor, foam polyethylene dielectric
C = Cablewave FXA78-50, smooth aluminum outer conductor, foam polyethylene dielectric
D = Cablewave FLC12-50, corrugated copper outer conductor, foam polyethylene dielectric
E = Cablewave HCC12-50, corrugated copper outer conductor, air dielectric

Table 5.1b. Laboratory direct shear calibration of grouted cables–
influence of shear zone width and grout voids (from Peterson, 1993).

Cable	Diameter (mm)	Linear Regression			Comment
		slope, S (mrho/mm)	x-intercept, u_o (mm)	r^2	
				Shear zone width	
F	19.1	14.9	1.7	0.9723	5 mm gap
		8.3	1.2	0.9696	20 mm gap
		9.9	1.6	0.9738	40 mm gap
		1.0	2.4	0.9606	80 mm gap
				Grout voids	
B	12.7	15.8		0.8888	no void
		7.7		0.9780	void @90
		24.2		0.8812	void @135
		9.8		0.9395	void @270
F	19.1	9.8		0.9455	void @90
		16.4		0.9190	void @135
		19.0		0.9566	void @180

Cables
B = Cablewave FXA12-50, smooth aluminum outer conductor, foam polyethylene dielectric
F = CommScope P3 75-750CA, smooth aluminum outer conductor, expanded polyethylene dielectric

Table 5.1c. Laboratory direct shear calibration of grouted cables—
influence of cable coating and distance from TDR pulser (from Kim, 1989).

Cable	Diameter (mm)	Linear Regression			Comment
		slope, S (mrho/mm)	x-intercept, u₀ (mm)	r²	
			Cable coating		
B	12.7	15.0	1.4	0.9579	coated
		13.7	2.1	0.9432	not coated
			Distance from tester		
B	12.7	15.1	1.4	0.9588	1 m
		11.5	1.6	0.9545	12 m
		10.9	1.8	0.9634	24 m
		6.7	1.8	0.9436	41 m
		4.7	1.9	0.9445	59 m
		10.7	1.8	0.9475	51 m, calculated
		4.5	1.7	0.9383	51 m, calculated
C	22.2	18.8	6.1	0.9471	1 m
		13.9	5.0	0.8694	51 m, calculated
		15.2	6.7	0.9313	51 m, calculated

Cables
B = Cablewave FXA12-50, smooth aluminum outer conductor, foam polyethylene dielectric
C = Cablewave FXA78-50, smooth aluminum outer conductor, foam polyethylene dielectric

Simple Extension

Laboratory tests in which grouted cables were subjected to simple extension have been performed by Su (1987) and Aimone-Martin et al. (1994). Su cut the grout (cured for 14 days) and pipe (but not the cable) into two segments and then produced extension by pulling apart the two pipe segments. The gradual separation between the two segments and the associated TDR reflection signatures were recorded in digital form until the cable either broke in tension or the bond between the cable and grout failed. Similar to the behavior observed during shear, the initial load-deformation behavior in tension is relatively stiff until yielding of the grout at the saw cut. The yield load is proportional to the cable diameter since it is controlled by the tensile strength of the cable and the cable-grout bond. Grout bond strength is discussed in Appendix A.

As shown in Figure 5-8a, the signatures generated during extension are much broader than those shown in Figure 5-8b for shear deformation. The width of the signature increased as the cable was extended until either the outer conductor or bond between grout and cable failed (i.e., pull-out occurred). Failure of the outer conductor is indicated in Figure 5-8a by an open circuit condition (large, positive reflection coefficient) at the breaking point. A TDR signature generated by shearing is much more easily detected than a signature generated by extension and the difference in grout-cable failure for these two modes is shown as inserts in Figure 5-8. For the 12.7 mm diameter cable tested, extension produced a negative reflection coefficient of approximately 20 mρ at failure. By comparison, a reflection coefficient of 150 mρ was obtained for shear failure at the same distance from the cable tester. Despite this low sensitivity in extension, cable failure due to extension can be distinguished from cable failure due to shear as shown by comparison of the termination reflections in Figure 5-8. Shear failure produces a large negative reflection as the outer conductor is brought into contact with the inner conductor. Extension failure produces only a large open circuit reflection as the inner and outer conductors remain concentric.

Aimone-Martin et al. (1994) conducted a series of tests with cables grouted into rock salt obtained from a site in New Mexico.

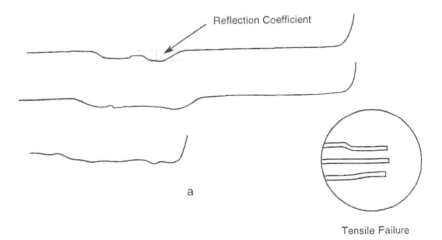

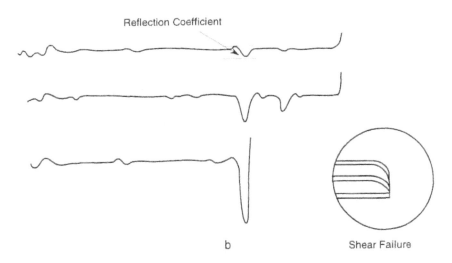

Figure 5-8. Comparison of cable outer conductor during failure in tension and shear ; (a) tensile failure; (b) shear failure (from Su, 1987).

The salt was cored 124 mm in diameter and cut to 330 mm lengths. A 32 mm diameter hole was drilled through the center of the core for grouting in the TDR cable. Two metal sleeves were epoxied at each end of test specimens. Two reference crimps were placed on each end to track sample extension. An increase in the distance separating the reflections was not detected until the rock failed in tension and the load was transferred from the rock to the cable. Corrugated copper cables provided useful separation data while smooth aluminum cable did not. Subsequent to rock failure, loads on aluminum cables did not significantly increase. This was attributed to cable slip along the grout-cable interface during tensile deformation. Slip was verified by post-test examination of the cables. Therefore, only corrugated cable types were used for correlation purposes. The results are discussed below in a comparison with combined loading.

Combined Shear and Extension

Combined shear and tension tests were performed by Aimone-Martin et al. (1994) using an apparatus shown in Figure 5-9. The apparatus was designed with the following components:

1.   381-mm-long structural tubes designed to carry and transfer the total load P to the cable and salt. Each tube has three holes, 95.3 mm apart, to accommodate 12.7-mm pin connections.
2.   The holes link members to (B) the end platens to vary the ratio of transverse shear to longitudinal tensile displacement depending upon the hole location used.
3.   Hollow circular sample sleeve jackets containing the salt-grout-cable sample. They are designed to slide vertically past one another to allow transverse shear of the cable and salt sample along the plane created by the support bars.
4.   Support bars (A) that transfer the shear load while preventing apparatus buckling.

During each stroke step, a total load P was placed on the testing apparatus. A portion of this total load was transferred as a shear force at the device steel jackets, causing the jackets to move past each other along the plane indicated by a thicker line (A) and inducing a localized shear in the salt-grout-cable sample. Simulta-

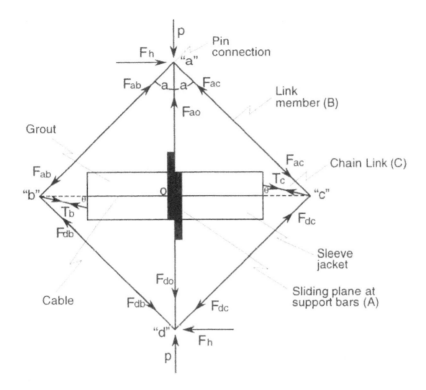

Figure 5-9. Free body diagram of combined shear/tension apparatus (from Aimone-Martin et al., 1994).

neously, the four link members (B) converted compressive force at the vertical axis to create tensile forces along the length of the chain link (C). The compression load P and chain link tensile loads T_c and T_b were monitored and recorded. Three unknown forces exist at pin "a" rendering a statically indeterminate problem. However, unknown forces within the link members at points "b" and "c" could be determined knowing T_b, T_c, and θ. At any stroke step, these forces could be determined as a function of θ and α. Thus, the unknown shear force F_{ao} at pin "a" and horizontal components of the chain tensile force could be calculated.

The types of cables sheared included helical corrugated

copper outer conductor with foam dielectric (Cablewave FLC12-50), smooth aluminum conductor with foam dielectric (FXA12-50), and helical corrugated copper outer conductor with air dielectric (HCC12-50). Digitized TDR waveforms were acquired at the end of each stroke step.

Eight combined loading tests were conducted on corrugated copper with air and foam dielectric cables with the outermost pin connection. One additional test was conducted with a 22.2 mm diameter smooth aluminum cable with a total shear deformation of 63.5 mm, producing extensive slip. Change in the distance between reflections associated with the reference crimps was negligible, while a shear deformation crimp did not develop. Based on these results and previous results in tension, smooth-wall aluminum outer conductor cables were excluded from further combined testing. However, these results are consistent with studies done by Su (1987), which demonstrated the need to have a sufficient bond length to prevent pull-out.

Table 5.2 gives a summary of calibration data for all tests. Data for combined shear and tension are compared with pure shear and pure tension. All data-fits are linear. Deformation plots indicate that air dielectric cables have a higher sensitivity in shear than the foam dielectric cables. In all cases of combined loading, failure occurred along the shear plane because of necking (extension) of the outer conductor, producing an open end as opposed to a short circuit. The ability to detect and track shear movement for combined loading was hampered by the broader shape in the shear waveform, as extension of the shear crimp prevented the rapid downward growth observed during pure shear.

The distance between TDR reflections associated with reference crimps was tracked for combined loading. Comparing the sensitivity for pure tension to that for combined loading shows that, essentially, a unit change in separation between TDR reference reflections was produced by a unit change in tensile displacement. However, for the air dielectric cable subjected to pure tension, it required 2.7 mm of displacement to create a 1.0 mm change in distance between the TDR reflections due to "smoothing out" of the corrugations.

Table 5.2. Laboratory calibration of grouted corrugated copper cables
(Aimone-Martin et al., 1994).

	Shear (mrho/mm)		Tension (mm/mm)	
	Pure shear	Combined loading	Pure tension	Combined loading
Air dielectric	25.6	3.8	2.7	1.0
Foam dielectric	13.9	2.3	1.2	0.8

Cable severance under combined loading required nearly twice the displacement compared with pure shear. The shear displacement range of both foam- and air-dielectric corrugated copper cables is similar for pure shear and for combined loading (i.e., the displacements at failure are 15.2 mm and 25.4 to 27.9 mm in pure shear and combined loading, respectively). Again, the tensile displacement range of air-dielectric cable is greater than that of foam-dielectric cable due to the air cable's higher degree of ductility. As cables extend, the helical corrugations smooth out, allowing a large range of tensile displacement prior to severance.

RESOLUTION AND ATTENUATION

Two phenomena were discussed in Chapter 2 that impact the assumption that cable deformities and their associated electrical discontinuities can be distinguished in time and space: 1) resolution of the distance between adjacent deformities, and 2) attenuation of reflected voltage pulses.

Resolution

The resolution of a measurement system is the smallest distance between two cable deformities or electrical discontinuities that can be measured before the discontinuities appear as one. It is dictated

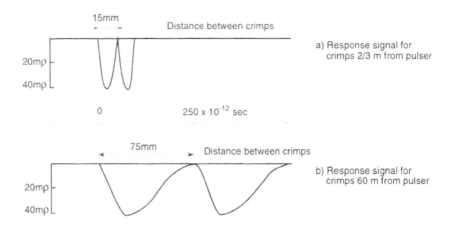

Figure 5-10. Influence of distance from TDR pulser on the resolution of TDR reflections (from Su, 1987).

primarily by the rise time of the voltage pulse arriving at the discontinuity and the timing capabilities of the TDR tester. Degradation (increase) of rise time of any pulse depends upon the cable attenuating the pulse.

The effect of rise time on the reflected pulse is illustrated in Figures 5-10 and 5-12. When multiple crimps or shears occur close to each other, the TDR system resolution will determine whether or not reflected spikes will overlap. In order to assess resolution using a Tektronix 1502B cable tester, cables were crimped at progressively smaller spacings until it was not possible to distinguish between the TDR reflection spikes. The resulting voltage spikes in Figure 5-10 were obtained for two cable crimps at distances of 0.6 m and 60 m from the cable tester. These tests indicate that a degradation in the rise time caused by dispersion altered the system resolution from 15 mm to 75 mm at 0.6 m and 60 m, respectively, using a TDR tester that launches a step voltage pulse with a rise time of 200 picoseconds.

Attenuation

Attenuation of a nondispersive, transverse electromagnetic (TEM) voltage pulse increases as frequency increases. This frequency effect for coaxial cable can be expressed as attenuation, α, in decibels per 30.5 m (100 ft) as a function of frequency (Mooijweer, 1971, and Table 10.1d),

$$\alpha = 10^{(0.523\log f) - 1.09}, \qquad\qquad (5\text{-}4)$$

where frequency, f, is in units of Hertz (cycles per second). Compare this relationship for attenuation along a coaxial transmission line with that for attenuation along an embedded parallel rod probe (Equation 2-35).

The attenuation effect for a nondispersive wave is calculated by deconvoluting the step pulse via a Fourier transform (Brigham, 1980) into a series of sinusoidal functions each with its own amplitude, frequency, and phase. The amplitude associated with each frequency is then reduced according to the attenuation relationship in Equation 5-4. The resultant waveform at any desired distance is then formed by combining the reduced voltage waveforms associated with each frequency via an inverse Fourier transform operation. Results of such a calculation show that there is only a 3.4% reduction in voltage and a 20% increase in rise time (time required for the voltage to increase from zero to its maximum) at 152 m for 12 mm diameter solid aluminum coaxial cable. Even at a distance of 1528 m, a nondispersive signal would be attenuated by only 29%.

TDR pulses are actually dispersive and the rise time increases substantially as the voltage pulse propagates along a cable. This degradation has a major influence on resolution (Taflove, 1995) as shown in Figure 5-10. There have been theories hypothesized for the effects of reflected pulse reduction, multiple reflections, or degradation of the rise time (Bartel et al., 1980).

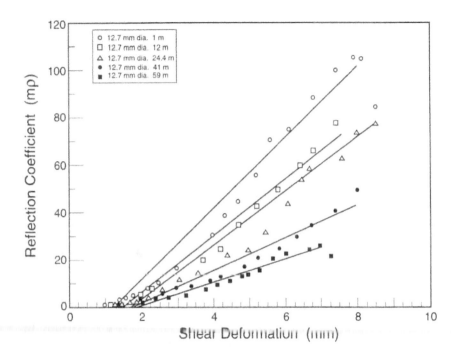

Figure 5-11. Attenuation of TDR reflection magnitude as distance from pulser increases (from Kim, 1989).

For distances less than 59 m, results of previous work (Kim, 1989) are combined in Figure 5-11 to show the degradation of reflected pulse amplitude with increasing length of cable. For example, shearing a 12.7 mm diameter cable by 6 mm produces a reflection coefficient of approximately 70 mρ at a distance of 1 m, 30 mρ at a distance of 41 m, and only 20 mρ at 59 m.

For distances greater than 60 m, tests were performed using a spool of 22.2 mm diameter solid aluminum coaxial cable (CommScope P-3 75-875CA) which was 530 m long (Pierce et al., 1994). Experimentation with a successively shorter length of cable was begun by cleanly cutting the end with a hacksaw. Shear and crimp deformations were then consistently applied along successively shorter cable lengths of 530, 268, and 94 m. Crimps were produced by squeezing the cable with a set of adjustable vise grips. The depth of the crimp is set by adjusting the separation of the

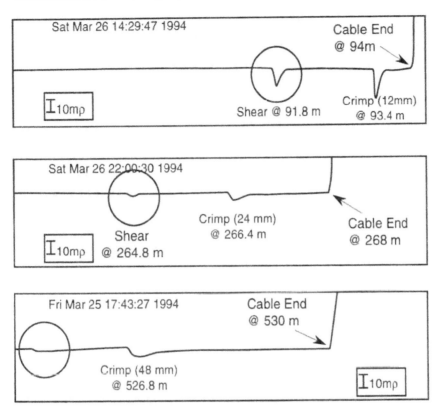

Figure 5-12. TDR reflections for a crimp and for a shear deformation of 7.6 mm: (a) 94 m long cable, (b) 268 m long cable, and (c) 530 m long cable (from Pierce et al., 1994).

grips. An optimally crimped cable diameter was found to be 15 mm, corresponding to a 30% diameter reduction. The length of the crimp was controlled by the grip width of 12 mm. Total crimp length was controlled by placing contiguous crimps in succession along the cable. Crimp lengths varied from 12 mm (1 grip width) for the shortest cable, as shown in Figure 5-12, to 48 mm (4 grip widths) for the longest cable. All reflections from these reference crimps were at least 5 mρ in amplitude.

Shear deformation was generated by a special device to produce a single shear in the cable (Pierce et al., 1994). The device is constructed in two halves to fit around the cable, with a screw to drive the inner shear block. The screw was manufactured with 20 threads per inch, or approximately 1.25 mm of shear deformation per rotation. The single localized shear produced by the device produced the TDR reflection highlighted with a circle in Figure 5-12.

A consistent testing procedure for crimp and shear deformation was employed near the end of each length of cable. The four-step procedure consisted of first placing a crimp within 3.3 m of the cable end, then shearing the cable in front (upstream) of the crimp, then shearing the cable again in front of the first shear, and finally placing a second crimp in front of the second shear. This routine yields a series of crimp, shear, shear, crimp, and open end discontinuities in that order, progressing toward the cable end, and is referred to as the "basic testing procedure."

Results are presented in three main sections: (1) influence of distance, (2) influence of multiple deformations, and (3) findings from crimp optimization and resolution. A summary of the data collected during this study is presented in Table 5.3 under the categories listed as undeformed cable, crimped cable, and sheared cable. The actual length of each cable was determined as the distance from the TDR tester to the cable end, marked by a near vertical upward reflection as shown in Figure 5-12.

The crimp lengths chosen for each cable and their corresponding reflection amplitudes are listed in the table. Resolution measurements were determined from two crimps separated by progressively shorter distances. Sensitivity results, and reflection amplitudes for a shear deformation of 7.6 mm, are displayed in the sheared cable column.

Influence of Distance

Distance has a dramatic effect on the signature produced by a single shear, as illustrated in Figure 5-12 and shown in Table 5.3. This figure compares the TDR reflections produced by shear deformation of 7.6 mm. Shear on the 94 m cable produced a typical spike with a 12.5 mρ amplitude reflection. At 268 and 530 m the shear reflections are more trough-like, with smaller amplitudes (2.5 and 1.2 mρ, respectively) and greater widths than the spike at 94 m. The reflection magnitude at 264.8 m is more easily distinguished

Table 5.3. Summary of results for undeformed, crimped, and sheared cables (Pierce et al., 1994).

Undeformed cable		Crimped cable			Sheared cable	
Cable length (m)	Open end reflection (mρ)	Crimp length (mm)	Reflection amplitude (mρ)	Resolution; minimum distance between crimps (mm)	Shear deformation required to produce 1 mρ reflection (mm)	Reflection amplitude at 7.6 mm shear deformation (mρ)
94	797	12	22	200	1.3	12.5
268	—	24	5	460	3.8	2.5
410	—	48	9	—	5.1	1.9
530	747	48	5	1520	7.6	1.2

than the one at 525.2 m, and can be identified as shear deformation. This suggests that small shear displacements will produce detectable reflections at distances up to 530 m, but the reflection magnitudes may not be sufficient to quantify movement. This insufficiency arises from the low sensitivity (smaller amplitude) and low resolution (larger reflection width) of reflections at this distance.

Figure 5-13 compares reflected amplitudes of 12.7, 10.2, 7.6, and 5.1 mm shear deformation at transmission distances ranging from 3 to 530 m. The family of curves for this range of shear deformation was obtained from the multiple measurements. Degradation occurs rapidly and minimum shear deformation necessary to detect a 1 mρ change in the waveform magnitude increases with distance from the cable tester (this is a function of the cable tester rise time as discussed in Hewlett-Packard, 1988). This reduction in amplitude results from attenuation of the signal with distance along the cable, as well as the degradation in rise time. It is important to realize that this chart is valid only for the bare CommScope 22.2 mm diameter cable used in the study. Other calibration charts should be developed for other types of cables with variable diameters and lengths to create a workbook of correlation plots. Adjustment must also be made for the initial displacement required to fracture grout encasing cables as shown in Figure 5-5 and Table 5.1.

Influence of Multiple Deformations

The effect of upstream crimps on existing shear and crimp reflections was minimal (Pierce et al., 1994). The general trend of minor degradation (less than 1 mρ) of reflections due to an upstream crimp was noted for all lengths of cable investigated. Additions to the basic procedure (crimp-shear-shear-crimp) were also examined to further define signal degradation. For the 410 and 530 m distances, a third crimp was placed at a greater distance up the cable from the second crimp. This third crimp also had virtually no effect on the existing signal reflections (i.e., difference < 1 mρ).

To investigate the impact of large shear deformations on TDR reflections, a third shear was placed a short distance upstream

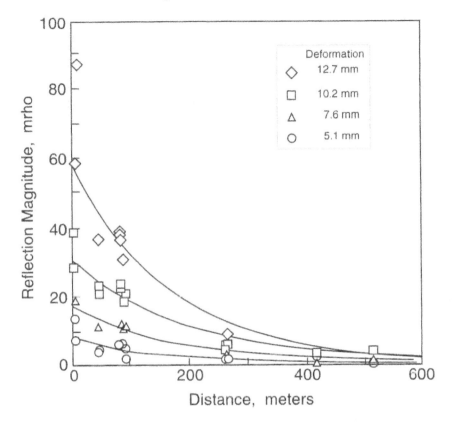

Figure 5-13. Influence of distance from TDR pulser on magnitude of reflection. Data shown for four different values of cable deformation (from Pierce et al., 1994).

of the second crimp and taken to failure, thereby producing a new open end at the shear location. This shear also produced minimal change in the magnitude of reflections made by previously existing crimps and shears.

Shear growth was tracked to failure. Amplitude increased with deformation until a short circuit was created when shear deformation was between 17.8 mm and 19.1 mm. Subsequently an open end occurred at a shear deformation of 27.5 mm, which is larger than the initial diameter of the cable. This response indicates that deformation can be correlated with the reflection coefficient to 78% reduction of the original diameter. However, the relationship between TDR reflection magnitude and cable deformation becomes highly nonlinear at such large deformations.

Crimp Optimization and Resolution

Table 5.3 (columns 3 and 4) lists the crimp length and respective amplitude for crimps produced on 94, 268, 410, and 530 m cables. At 94 m, which is a typical cable length in present field applications, a single crimp, 12 mm wide and 15 mm deep, yields a 22 mρ spike. This reflection, shown in Figure 5-12, is easily detectable with TDR and serves as an excellent location marker. Even at 530 m, a 48 mm crimp (with no upstream deformation) produces a partial spike with an amplitude of 5 mρ, which is still discernible and useful for marking a location along the cable. Therefore, cable lengths up to 530 m can be accurately marked with relatively little deformation.

As the cable length increases, the crimps must necessarily be longer for identification, and a combination of these two factors produces signal reflections that become correspondingly wider. Thus it is important to minimize the number of crimps and their dimensions, especially at greater lengths, to produce a small, yet discernible voltage reflection. Current field practices involve addition of multiple reference crimps yielding 20 mρ or more of reflection. However, the results in Table 5.3 indicate that crimps with 5 mρ amplitude would be sufficient for marking cables, thereby reducing the required crimp length.

The number of crimps should be minimized to reduce the amount of energy loss that occurs as the pulse passes through these multiple deformations (e.g., see Figure 6-1). A number of options are available. If the location of interest is known, a single crimp may be placed near this location. Perhaps two crimps might be employed if the location of deformation is dispersed or unknown. Unless very accurate locations of cable defects are necessary, a small number of crimps are recommended. Measurements presented herein show that one and two crimps will not affect the TDR reflection magnitude caused by cable deformation in response to rock or soil movement.

Resolution clearly decreases in a nonlinear manner as distance from the source to the crimps increases, as indicated by the data in Table 5.3 (column 5). An inverse relationship between resolution and distance should be expected as the rise time increases with cable length and subsequently increases the width of signal reflections. This trend is demonstrated in Figure 5-12, where the apparent width of crimp reflections increases from 0.6 m to almost 2 m as the cable distance increases. Since the actual crimp lengths increase with cable length, some of this increase reflects the increase in actual crimp as well as the width of the reflected signal. The crimp reflection widths are also larger than the corresponding shear reflection widths in each case, which indicates that resolution is governed by the crimps. Thus two adjacent shears can be resolved at distances smaller than those listed in the table based on crimp resolution. Therefore, resolution values in the table are conservative.

NOISE

For long cables, noise is also an important consideration, which can be deduced from Figure 5-12. For example, the noise level was typically 1 to 2 mρ in magnitude for the 530 m cable. Since a 10 mm shear produces a 4 to 5 mρ reflection amplitude at this distance, TDR is limited in quantification of deformation to larger shearing events with long cables. However, as shown in Figure 5-13, quantitative measurement of small shear displacements is still possible for cables as long as 268 m. Connecting a cable tester to a transducer cable with a low-loss lead cable may allow quantitative measurement at distances greater than 268 m.

Random noise is a problem which has arisen on several field projects in which remote monitoring was used to acquire waveforms from a coaxial cable installed in a hole drilled from the surface down to an underground coal mine. The noise amplitude was approximately 50 mρ while the magnitude of reference crimp reflections was only 10 mρ. This noise can be random in time and location of occurrence. At other projects, it has appeared along the entire length of a cable and was so severe that it was not possible to detect TDR reflections. Very often the problem can be rectified by ensuring that only good quality shielded cable is used between the cable tester and the downhole transducer cable. The shield must be properly grounded (Cablewave Systems, 1985). Research to determine cause(s) of noise and to develop techniques which can be used to eliminate it from the TDR waveforms is continuing.

Chapter 6
FIELD EXPERIENCE AND VERIFICATION OF ROCK DEFORMATION MEASUREMENT

In situ measurement of rock deformation is compared in this chapter with TDR monitoring of cable deformation in a number of differing geologic settings and deformation modes. To date several dozen TDR cable measurement systems have been installed worldwide, which underscores its growing acceptance in the mining and geotechnical communities. A number of comparisons are presented to demonstrate diverse applications of quantitative evaluation of TDR waveforms obtained from cables installed in rock masses undergoing large deformations. Comparisons are made with direct measurement of shear displacement in shallow boreholes and inclinometer measurements in deep boreholes. Indirect comparisons based upon beam bending and cumulative horizontal displacements are also presented. This chapter closes with a number of case histories that illustrate the variety of environments wherein TDR measurement has been employed successfully.

COMPARISON WITH INCLINOMETER

Two 200 m TDR cables were installed within the rock and soil overburden above a longwall coal mine panel, as shown in Figure 6-1, to monitor ground behavior associated with subsidence (Mehnert et al., 1992; Kawamura et al., 1994). The surficial materials were composed of 3 to 6 m of Illinoisan glacial drift overlain by the 1 to 2 m thick Bluford silt loam. The bedrock overburden included 213 m of Pennsylvanian strata, consisting of 35% shale, 37% siltstone, and 18% sandstone above the Herrin

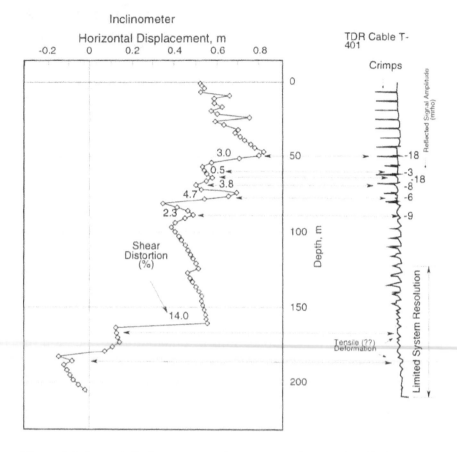

Figure 6-1. Lateral displacements at the center of a longwall coal mine panel when mine face was 18 m from instrumented boreholes (from Kawamura et al., 1994).

(No. 6) Coal Seam which was located about 220 m below the ground surface, with an average thickness of 3.2 m at the instrumented site. The mine panel was 183 m wide and 1524 m long.

One cable was installed in the dynamic subsidence zone at the panel centerline (T-401) and a second cable was installed 61.3 m from the panel centerline (T-400) in the static tensile zone. In addition, an inclinometer/sondex system was installed at the panel centerline about 4.8 m west of T-401. Both cables were 12.7 mm diameter, unjacketed, Cablewave FXA12-50. Reference crimps were placed at intervals of 6.1 m to provide accurate distance measurements. As many adjacent crimps as necessary to obtain a desired reflection magnitude of 40 mρ were placed as shown in Figure 6-1.

The cables were installed in 76 mm diameter boreholes using a grout with a 65% water-to-cement ratio (high early strength Type III cement) and 2% by weight of an expansive agent (Chemcomp III) up to the top of bedrock. Cable T-401 was grouted to within 3 m of the soil-rock interface (depth of 8 m), above which it was backfilled with bentonite grout in order to provide a stiffness comparable to the surrounding ground. Cable T-400, 213.1 m long, was grouted from the bottom up to a depth of 48.8 m. The remainder of the borehole was backfilled with bentonite grout.

The 207.3 m long inclinometer/sondex system was installed in a 203 mm diameter borehole. The system consisted of an 85 mm diameter ABS plastic inclinometer casing which was placed inside a 101 mm diameter corrugated, flexible plastic casing with sensing rings. The sensing rings were spaced every 6.1 m from the ground surface down to 140.2 m, and every 3 m from 143 m to the bottom of the casing. The 51 mm annulus between the borehole wall and sondex casing was grouted with a bentonite slurry up to the ground surface.

The measurements obtained by cable T-401 were generally consistent with the readings taken with the inclinometer/sondex system (Figure 6-1), which indicate that shear distortions of 0.5% to 14% occurred when the face was about 18 m away. Several reflection spikes, which ranged in magnitude from 4 mρ to 18 mρ, were measured at comparable depths. The location of the reflection spikes and shear distortions were invariably associated with: 1) the presence of presubsidence fractures, or 2) contact between units with contrasting stiffness. Shear reflection spikes as high as 58 mρ (or 3.3 mm using the calibration presented in Figure 5-5b) had been previously recorded at other sites. Therefore, the last set of readings were taken when the cable was still distant from the failure condition and additional data could have been obtained if a remote monitoring capability had been used.

Table 6.1 summarizes the correspondence between the inclinometer and TDR readings. Since the readings were recorded using a strip chart, it was not possible to obtain a reliable time history of cable deformation. Two major shear distortions with

Table 6.1. Comparison of inclinometer displacement and TDR reflection magnitude (after Kawamura et al., 1994).

Depth	Inclinometer		TDR		Ratio of TDR disp. to inclinometer displacement (mρ/mm)
(m)	Slope 0.6 m probe (%)	Displacement (mm)	Reflection magnitude (mρ)	Shear displacement[1] (mm)	
49.7	3.0	18	18	5	0.25
59.7	0.5	3	5	3	0.95
61.9	1.4	8	18	6	0.70
67.7	3.8	23	8	4	0.17
77.4	4.7	28	7	4	0.13
89.0	2.3	14	9	4	0.31
160.0	14.0	84		tension ?	
180.0	7.0	42		tension ?	

[1] Computed using linear regression from Table 5.1c (Distance from tester, Kim, 1989)

inclinometer magnitudes of 7% and 14%, which were below a depth of 150 m, were not detected by TDR monitoring of cable T-401. At these positions, it is likely the TDR reflections associated with cable deformation overlapped with reflections from reference crimps made in the cable, and the cable tester gain was not set high enough to detect these reflections.

Cable T-401 was sheared off at a depth 35.1 m, when the mine face was 30 m past the cable. The shear break was located within Mount Caramel sandstone which is interbedded with thin silty shale. Measurements made with the Sondex indicated a small differential vertical extension of 2.3 mm between the depths of 32.1 m and 37.9 m, when the face was located 14.9 m past; the corresponding average vertical strain was 0.04%. Furthermore, a post-subsidence horizontal, uneven to rough, micaceous fracture at a depth of 35.2 m was observed in a post-subsidence borehole located 60 m from T-401. Several TDR reflections, which ranged in magnitude from 1 mρ to 48 mρ, were measured as fracture propagation continued over a period of 60 days. These reflections, due to cable deformation, were apparently associated with slickensides that developed within the rock mass as a consequence of subsidence and were observed in post-subsidence core samples.

Inclinometer/sondex casing cannot accommodate localized shear displacements and distortions. This deformation squeezes the casing, which blocks the downward movement of the inclinometer and sondex probes through the 85 mm diameter casing. The sondex probe (0.3 m long) is less affected by distortions than the inclinometer probe which is about 0.6 m long. Both probes were blocked at a depth of 43.8 m when the mine face was 66.4 m past. A subsequent set of readings, taken when the face had progressed to a distance of 194 m past, indicated that the inclinometer probe was again blocked at that depth, while the shorter sondex probe was able to penetrate more deeply. Both of the probes crossed several locations along the inclinometer casing with alignment deflections as high as 35 degrees (slope of 61%) before being blocked. Further discussion of, and comparison with, inclinometer measurements is presented in Chapter 7.

COMPARISON WITH OBSERVATION HOLES

In 1989, TDR was introduced for geomechanical monitoring of the performance of two 91-m long by 10-m wide by 4-m high test rooms of the Waste Isolation Pilot Project (WIPP) (Francke et al., 1994). These rooms were excavated in salt within the Permian Salado Formation. Some 4 m and 2 m above the room are located 21 cm and 6 cm thick clay seams that provide weaknesses in the roof. Cables were installed to measure movement along these seams using TDR.

Three roof boreholes approximately 3.7 m deep and 7.6 cm diameter were drilled at the midpoint of the two test rooms (Rooms 2 and 4). The 12.7 mm diameter, unjacketed coaxial cables (Cablewave FXA 12-50) were cut to 4.3 m lengths and installed using procedures similar to those used previously to monitor behavior of roof strata in a potash mine (O'Connor and Zimmerly, 1991). Prior to installation, a UHF connector was attached to the end of the coaxial cable and connected to a cable tester. The coaxial cable was then crimped at 0.6 m intervals. Crimping was monitored with the cable tester to ensure that the crimps provided a strong reflection (approximately 20 mρ) but did not short circuit the cable. Upon completion of crimping, each cable was attached to an equal length of 5 mm diameter vent tube and inserted into the borehole. A grout hose was inserted approximately 15 cm into the borehole and the entry to the borehole was sealed. The cables were grouted into the borehole using an expansive cement grout as shown in Figure 6-2.

The cables were installed in holes located adjacent to roof observation holes. The observation boreholes were inspected using an aluminum probe rod with a flattened nail or screw about 1.6 mm wide attached to one end. Features along the observation hole were identified by scratching the nail along the sides of the borehole while applying moderate pressure. Features were noted when the nail caught on the borehole wall at the same depth on all sides of the wall. Strata separation greater than 4.8 mm was determined by the amount of vertical movement within the feature. Horizontal displacement magnitude was visually estimated when possible.

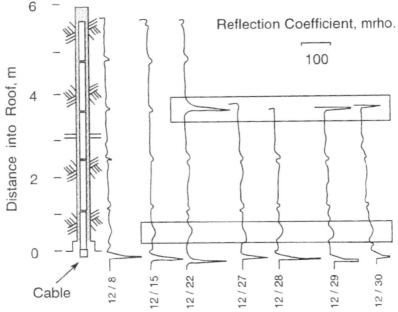

Figure 6-2. Roof installation of cable at potash mine and TDR waveforms (from O'Connor and Zimmerly, 1991).

When visual estimation was not possible, the magnitude was estimated by feel (with the rod). Accuracy of this technique was ± 3.2 mm (Francke and Terrill, 1993).

In order to correlate TDR reflection magnitude with displacement, a laboratory double shear test similar to that illustrated in Figure 5-2 was performed on grouted cables. To obtain a correlation between reflection amplitude and shear displacement, linear, multiplicative, and exponential regressions were performed. The exponential correlation,

$$u=[1n(V_{pk}+100)-a_1] \; / \; a_2 \qquad (6\text{-}1)$$

was selected where u is the shear displacement in inches, V_{pk} is the TDR reflection amplitude in mρ, and the regression constants a_1 and a_2 are 4.43 and 1.95, respectively. Correlation coefficients ranged from 39% to 97% for 12 data sets. Two sets of data with suspect measurements were deleted, resulting in an overall correlation of 90% and an overall standard deviation of 0.17 mρ/mm (4.26

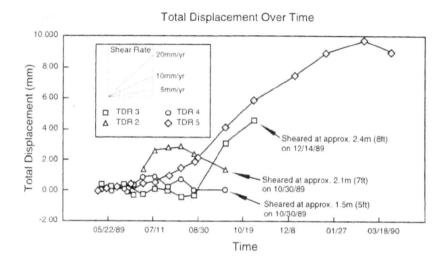

Figure 6-3. Time histories of displacement at a depth of 2.4 m in roof of room excavated in salt (from Francke et al., 1994).

mρ/inch) for the remaining ten sets of data.

The comparison of TDR data to that from observation boreholes focuses on the clay seam at a depth of 2.4 m. A plot of total displacement over time for each TDR cable at 2.4 m is shown in Figure 6-3 and the results are summarized in Table 6.2. The annualized shear rates are based upon data for specific time periods rather than over the useful life of the cable. The apparent decrease in displacement implied by the time history plots in Figure 6-2 is discussed in the comparison with elastic beam theory later in this chapter.

TDR2 was installed on June 15, 1989, and data was collected regularly until August 11. Using the laboratory calibration (Equation 6-1), the shear displacement rate at a depth of 2.4 m was calculated to be 19.5 mm/yr. Complete shearing of the cable was detected on October 30. During the useful life of TDR2, the observation boreholes in the area were inaccessible and therefore, no observation borehole data were available for comparison. Due to deterioration of ground conditions in Room 2, access to the room was restricted. Prior to the restricted access, a 45 m

Table 6.2. Comparison with observation holes in
mine roof (Francke et al., 1994).

	TDR2	TDR3	TDR4	TDR5
Location	Room 2	Room 4	Room 4	Room 4
Date installed	6/15/89	5/11/89	5/12/89	5/10/89
Last date read before sheared off	8/11/89	10/30/89	9/29/89	3/29/90
TDR displacement (mm)	2.8	4.5	6.2	8.9
TDR shear rate (mm/yr)	19.5	9.6	16.6	10.1
Observation hole shear rate (mm/yr)	6 to 13*	7.9	9.5	7.9
Ratio of TDR shear rate to nearest observation hole shear rate	N/A	1.22	1.75	1.28
Distance (m) between TDR cable and nearest observation hole	N/A	2.5	2.1	2.6

*estimated

extension was attached to the cables to allow remote monitoring.
Before the room was closed, boreholes in the area had offset rates
between 6 and 13 mm/year.

TDR3 was installed on May 11, 1989 in Room 4 and was
found completely sheared at 2.4 m 216 days later on December 14.
Prior to failure, the shear displacement rate was 9.6 mm/yr, and the
nearby observation borehole indicated a shear rate of 7.9 mm/yr.
This difference in shear rates can be largely attributed to error in the
observation borehole shear measurement. The observation borehole
measurement, which is visually estimated, has an accuracy and
resolution of about ±3 mm. In addition, it has been determined that
the shear rate is highly dependent upon the location of the borehole
in the room cross section. Typically, shear displacement rates are

highest about 2 m from the ribs and decrease rapidly as the centerline is approached. Moving the measurement point about 0.5 m can reduce the shear rate by 50%.

TDR4 was installed on May 12, 1989 in Room 4 and found completely sheared 165 days later on October 30. Prior to failure, the shear displacement rate was 16.6 mm/yr and the nearby observation borehole indicated a shear rate of 9.5 mm/yr. The TDR cable shear rate was 75% higher than the shear rate observed in the observation borehole. This difference can also be attributed to the problems with the accuracy of the visual observation borehole measurements.

TDR5 had a shear rate of 10.1 mm/yr, whereas the shear rate in the nearby observation borehole was 7.9 mm/yr, a difference of 28%. This is within the accuracy of the borehole observation technique. Given that the observation boreholes were typically located approximately 3 m from the coaxial cables, the TDR shear rate correlated with borehole observation rates to within ±4.3 mm/yr. In addition, the TDR measurements correlated well with nearby observation boreholes in locating shear movement within the rock.

More recent applications of TDR at the WIPP facility have involved the use of 22 mm diameter corrugated copper air dielectric coaxial cable (Cablewave HCC 78-50). This has increased the sensitivity to shear deformation as indicated in Table 5.2. Aimone-Martin and Francke (1997) presented different correlations between shear displacement, u, and TDR reflection magnitude, V_{pk}, such as the inverted sigmoid

$$u = -a_1 \ln \left[\left(a_2 / (V_{pk} - a_3) \right) - 1 \right] + a_4 \qquad (6\text{-}2)$$

which they have fit to both laboratory and field data with the regression constants a_1, a_2, a_3, and a_4 having values of 0.15, 485.99, –6.83, and 0.70, respectively.

COMPARISON WITH ELASTIC BEAM THEORY

The photograph in Figure 6-4 shows installation of TDR cables to monitor roof strata movements during pillar extraction in a potash mine (O'Connor and Zimmerly, 1991). The study was initiated in response to a decision to implement secondary mining of pillars in order to extend the mine life. The langbeinite being mined is less plastic than the sylvite which is more commonly mined in the Carlsbad Potash Basin, and it was not certain that previous experience with pillar recovery at other mines would be applicable. Therefore, prototype secondary mining was conducted in one mine panel to assess mining equipment and procedures, and to determine if ground control problems would develop. An instrumentation array was installed to monitor strata movements, changes in pillar stress, and surface subsidence. The pillar extraction program was conducted from December 5, 1988 to January 31, 1989.

The mine is at a depth of 380 m in the 4th ore zone of the McNutt Potash Member where langbeinite is extracted using conventional room and pillar techniques. The overburden consists of approximately 9 m of Quaternary caliche, sand, and gravel, 30 m of siltstone and fine-grained sandstone (Dewey Lake Red Beds), 98 m of dolomite, anhydrite, gypsum, and siltstone (Rustler Formation), and 244 m of halite with thin beds of anhydrite and polyhalite (Upper Salado Formation). Clay seams are common at the base of these anhydrite and polyhalite beds, and the immediate mine roof consists of halite layers separated by clay partings.

A mine-level instrumentation array was installed to monitor roof–floor convergence, roof strata separation, and changes in pillar stress. Roof–floor convergence was monitored at 51 locations with a Philadelphia rod to measure the distance between spads driven 2.5 cm into the roof and floor. To monitor roof strata separation, 12.7 mm diameter coaxial cables (Cablewave FXA 12-50) were crimped at 1.2 m intervals and grouted into roof bolt holes 6 m deep at 15 of these locations.

In order to compare the roof–floor convergence measurement with TDR measurement of cable deformation, it was assumed that convergence primarily resulted from roof deflection with relatively little floor heave or squeezing of unmined pillars.

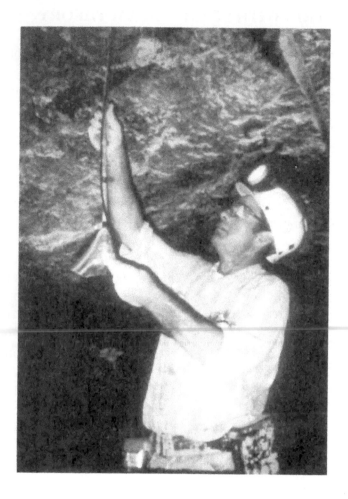

Figure 6-4. Installing coaxial cable in roof of a potash mine.

Convergence profiles could then be assumed to represent deflected beam profiles of the immediate roof strata. Furthermore, it was assumed that plane sections through a beam taken normal to its axis remained plane as the beam deflected. Under these assumptions, horizontal displacement of a point is linearly proportional to its distance from the neutral axis of the sagging roof beam.

There is a clay parting in the roof at a depth of approximately 3.6 m and shearing of the coaxial cable at this depth is

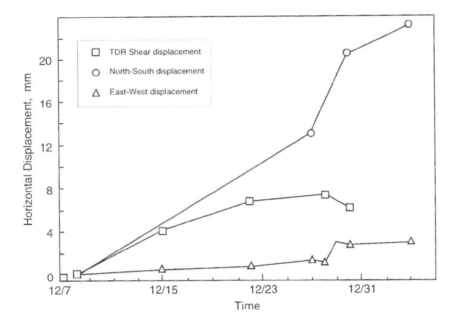

Figure 6-5. Time history of displacement at top of roof beam and also from TDR monitoring of cable deformation (from O'Connor and Zimmerly, 1991).

apparent in the TDR signatures in Figure 6-2. Treating this depth as the deflected beam thickness and assuming a prismatic beam cross section, the neutral axis is located at a depth of 1.8 m. For extreme fibers along the top of this beam, horizontal displacement is the product of 1.8 m times the slope of the deflection profile.

The convergence contours indicate that the roof deformed into a bowl. This mode of deformation implies that points in the immediate roof experience two horizontal components of displacement in addition to the vertical component. Computed components of horizontal displacement at a depth of 3.6 m are plotted versus time in Figure 6-5. The north–south component is consistently greater than the east–west component reflecting the asymmetrical shape of the convergence bowl.

TDR reflection magnitudes were converted to shear displacement from the correlation in Figure 5-5b. Shear displacement of the cables is consistent with the north–south component of horizontal displacement. The magnitude of TDR shear displacement attained a peak value and then decreased as the face advanced past the entry on 12/21/88. This decrease in magnitude of the TDR reflection in Figure 6-5 is actually associated with tensile deformation of the coaxial cable and has been observed during laboratory calibration tests (Su, 1987). This tensile deformation occurred as the immediate roof strata separated from the overlying main roof along the clay parting.

The distinct TDR reflection at a depth of 3.6 m in Figure 6-2 is associated with localized shearing of the cable. The trough-like reflection at a depth of 0.6 m corresponds with tensile deformation of the cable as strata separation developed along a clay parting. This separation could be observed at one location and it could be monitored by direct measurement. Its aperture increased from 25 mm to 112 mm during the period December 20–29, 1988. It is characteristic of cable extension that the TDR reflection increased in width but not in magnitude (compare Figure 5-8). This interpretation of TDR signature changes infers a scenario in which both the immediate roof and main roof initially cantilevered as a laminated beam without strata separation. After pillars on both sides of the cable location had been mined on 12/21/88, yielding of the remnant pillars accelerated the strata separation and convergence rate.

INSTALLATION DETAILS

Cables have been installed in holes of various orientations and depths. The most common hole orientation has been vertically down, and most holes have been less than 100 m deep. These installations have encountered the same problems inherent with all borehole instrumentation.

Deep Holes

When placing cables in holes deeper than 50 m, the following problems have occurred:

- Hole collapses have prevented passage of cable.

- Clays and shales have squeezed into an uncased hole and prevented passage of cable.
- Cables could not be advanced past protrusions in the borehole.
- Skin friction between the cable and borehole walls made it impossible to lower the cable.

Obviously, cable installation is not possible if a stable hole cannot be maintained. Typically, holes are stabilized with casing which is removed after the grout has been placed. It is best to minimize the time between completion of drilling and placement of the cable and grout.

When placing a cable into holes greater than 200 m deep, it has been necessary to place a 10 Kg (25 lb) weight at the end of the cable in order to overcome skin resistance and to provide a rigid, straight section that can be maneuvered past downhole protrusions. Another successful approach has been to attach the cable to flush-coupled PVC grout pipe as the pipe was lowered down the hole, then leaving the pipe in place after grout placement was completed.

Up Holes (mine roof)

O'Connor and Zimmerly (1991) and Francke et al. (1994) installed 12.7 mm diameter cables into 50 mm diameter roof bolt holes which had been drilled vertically up into a mine roof to a depth of 6 m. The cable was crimped at 1.2 m intervals. A 5 mm diameter bleeder tube was attached and the assembly was pushed up into the hole and held in placed with wire. Then a 0.15 m length of 12 mm diameter grout hose was placed at the bottom of the hole and it was packed off with oakum. Grout was pumped into the hole until it returned through the bleeder hose.

Panek and Tesch (1981) installed 12.7 mm diameter cables into 47.5 mm diameter holes which had been drilled 30° to 60° above the horizontal to depths of 42 to 188 m. They set a hydraulic-anchor-pulley device which was pushed up the hole with 12 mm diameter PVC water pipe. Hydraulic fluid was pumped up nylon tubing to expand the anchor. The coaxial cable with attached bleed tube was pulled up the hole with parachute cord passing over the pulley anchored at the end of the hole. A driller's piston-type water pump was then employed to pump grout very slowly into the hole through a pipe fitting at the sealed hole collar until grout emerged from the bleed tube.

APPLICATIONS

A variety of situations in which TDR has been used to monitor rock movement are listed in Table 6.3. The list is not exhaustive, but does demonstrate the diversity of TDR applications. Details are provided in the references, but some examples can be briefly summarized to illustrate the evolution of this technology as well as the diverse schemes in which it has been applied.

Continuous Monitoring—Strata Movement Over Active Mines

The study reported here was made at a longwall coal mine near West Frankfort, Illinois as the mining face advanced toward a cable installed in a hole drilled from the surface down to mine level (Dowding and Huang, 1994a). Longwall mining produces immediate planned subsidence over a panel. As the face advances and all the coal across the panel width is removed, the overburden behind the shields supporting the face is left unsupported and the immediate roof collapses. Vertical and horizontal ground movements propagate up through the overburden, and surface subsidence occurs. As shown in Figure 6-6, the TDR borehole was located at the center of the panel and the location of the advancing longwall face is indicated by the dated lines oriented north–south. TDR waveforms were acquired twice daily and compared with movement measured on the surface. Subsurface movements measured by TDR were shown to be localized and predominately shearing in nature. Localized changes were detected at depth some 4 days before movement was observed at the surface.

The overlying cyclotherm geology is typical of the Illinois Coal Basin. Coal seams at a depths of 38, 107, and 115 m are overlain with 8, 16, and 8 m of shales, respectively. The 0.2 m of

Table 6.3. Locations where TDR has been been used to monitor rock deformation in mines.

Location	Cable length (m)	Type	Application	Reference
Sesser, IL	200	A	high extraction coal	Pulse (1970)
San Manuel, AZ	42 to 187	A	block cave copper	Panek and Tesch (1981)
Benton. IL	189	B	high extraction coal	Dames and Moore (1983), Wade and Conroy (1980)
Weston, CO	270	D	high extraction coal	O'Rourke et al. (1982)
KY	55	D	abandoned coal	Dowding (1983)
Montcoal, WV	183	D	retreat pillar coal	O'Connor et al. (1983)
Marion, IL	75	A	high extraction coal	Mehnert et al. (1992)
PA	217	D	high extraction coal	Su and Hasenfus (1987)
Carlsbad, NM	6	A	retreat pillar potash	O'Connor and Zimmerly (1991)
Collinsville, IL	43	A	abandoned coal	Bauer et al. (1991)
Pecos, TX	140	B	in situ sulphur	O'Connor (1989)
Galatia, IL	122	A	high extraction coal	Van Roosendaal et al. (1992)
West Frankfort, IL	159	C	high extraction coal	Dowding and Huang (1994a)

Cables
A = FXA12-50, 12.7 mm, solid aluminum, Cablewave Systems, New Haven, CT
B = FXA78-50, 22.2 mm, solid aluminum, Cablewave Systems, New Haven, CT
C = P3 75-875 CA, 19.05 mm, solid aluminum, Comm/Scope, Catawba, NC
D = RG8AU, 12.7 mm, braided copper, Radio Shack, Chicago, IL

Table 6.3. (continued)

Location	Cable length (m)	Type	Application	Reference
Ft. McMurray, AB	53	B	tailings slope	O'Connor (1991), Peterson (1993)
	142	A	highwall slope	
	49 to 136	C	highwall slope	
Sesser, IL	224	A	high extraction coal	Brutcher et al. (1990), Van Roosendaal et al. (1990)
Goldenville, NS	7 to 26	A	abandoned gold	Hill (1993)
Empire, CO	431 to 609	–	block cave molybdenum	Doepken (1991)
Emery County, UT	463	C	high extraction coal	O'Connor (1992)
Cobalt, ON	5 to 23	A	abandoned gold	Charette (1993a, 1993b)
Benton, IL	100	C	high extraction coal	Siekmeier et al. (1992)
Timmons, ON	9 to 29	A	abandoned gold	Aston and Charette (1993)
Oakland, MD	189	C	high extraction coal	O'Connor (1994)
Collinsville, IL	61	C	abandoned coal	O'Connor (1995)
Cambridge, OH	12 to 15	C	abandoned coal	

Cables
A = FXA12-50, 12.7 mm, solid aluminum, Cablewave Systems, New Haven, CT
B = FXA78-50, 22.2 mm, solid aluminum, Cablewave Systems, New Haven, CT
C = P3 75-875 CA, 19.05 mm, solid aluminum, Comm/Scope, Catawba, NC
D = RG8AU, 12.7 mm, braided copper, Radio Shack, Chicago, IL

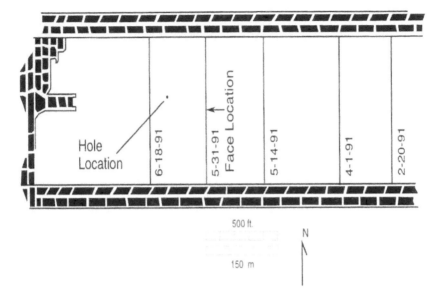

Figure 6-6. Location of cable installation over longwall coal mine panel (from Dowding and Huang, 1994a).

limestone at 30 m, 0.7 m of interlaminated siltstone and sandstone at 40 m, and 0.1 m of limestone at 66 m also provide significant changes in stiffness which localize deformation. The rock mass stiffness versus depth (Siekmeier and O'Connor, 1994) is shown in Figure 6-7.

Figure 6-7 presents TDR records obtained as the face advanced towards (–) and past (+) the cable location. Reflections highlighted with a "+" correspond with reference crimps used to improve location accuracy of measurements. Waveforms 1, 2, 3, and 4 were recorded before and after complete severance at depth of 66 m. In a similar fashion, waveforms 2, 3, 4, and 5 were recorded before and after complete severance at depth of 48 m. These records show that strata movement sheared the cable, producing the reflections such as those marked as (a), which increased in amplitude as the longwall face advanced. At depths of 48 m and 66 m, strata movement was sufficient to sever the cable, which produced the open end reflections marked as (b) and (c), respectively.

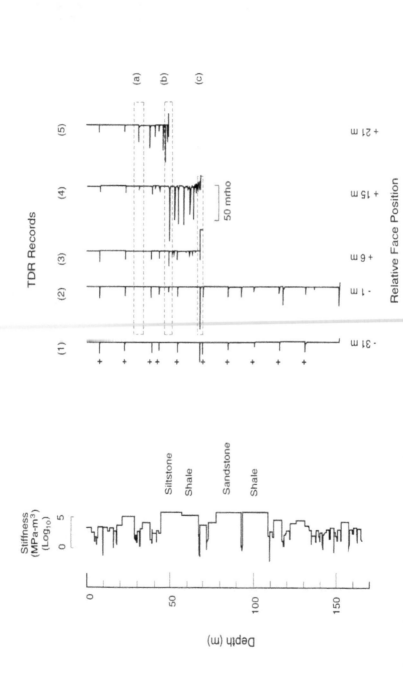

Figure 6-7. Rock mass stiffness profile and TDR waveforms for cable installed over longwall mine; reflections highlighted with a "+" correspond with reference crimps (from Siekmeier and O'Connor, 1994).

The correlation in Figure 5-7 and Table 5.1 was used to convert TDR reflection amplitude to rock strata movements at the depths of 48 m and 66 m. Cable length was considered to have the most influence on signal interpretation. If the shear zone thickness is less than 40 mm, the magnitude of the thickness has little affect on the correlation between reflection amplitude and shear displacement (Figure 5-7 and Table 4.1). However, if voids are present in the grout and are aligned parallel to the direction of shearing, they will reduce the sensitivity by some 50% (Table 5.1). Therefore, the presence of grout voids would reduce the sensitivity shown in Figure 5-7, as does the larger cable used.

Horizontal displacements at the surface are compared with subsurface shear displacement at depths of 48 m and 66 m in Figure 6-8. Horizontal displacement at the surface was measured by electronic survey of surface monuments (Bennett et al., 1992). Even more important than the time lag between displacement at depth and at the surface is the distance at which early subsurface detection occurs. Movements at a depth of 66 m began when the face was between 125 m and 64 m from the cable whereas they did not begin at a depth of 48 m until the face was 23 m from the cable. Surface movement was not detected until the face was within 15 m.

It should be recognized that the surface displacements are absolute values while shear deformations along the cable are relative values. However, hypothetically speaking, horizontal displacement on the surface at a particular time or distance to the face location should be equal to the sum of all subsurface shear deformations. Assuming that the direction of shearing deformation underground is additive and that it occurs in the same direction, a comparison can be made at an instant when the distance between face and cable was some 3 m, which was the last time for which complete cable and surface survey data are available. As can be seen in Table 6.4, summation of shearing deformations along the cable was 32.47 mm and 55.41 mm, respectively, depending upon what assumptions are made relative to the sensitivity as discussed above. By comparison, the horizontal displacement of the surface monument was 49.5 mm. While the agreement in these measures of displacement may be fortuitous, it nevertheless lends credibility to the ability to measure relative displacement with TDR reflection magnitude.

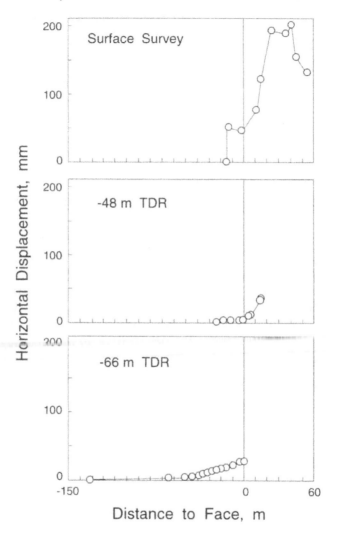

Figure 6-8. Comparison of horizontal surface displacement and cable shearing deformation at depth. Distance
between hole location and mine face shows early detection
of movement at depth by TDR (from Dowding and
Huang, 1994).

These field measurements verify the use of TDR measurement as an early warning system to detect subsidence before it
occurs on the surface. Comparison of TDR reflection magnitude
and surface subsidence measurements showed that localized subsur-

Table 6.4. Minimum and maximum estimates of
cumulative shear deformation along cable
when mine face was 3 m away from cable.

Depth	Reflection coefficient	Shear displacement (mm)	
(m)	(mρ)	minimum	maximum
48	8.96	3.8	5.7
66	65.30	15.8	29.6
91	12.80	4.6	7.3
115	10.88	4.2	6.5
148	10.24	4.1	6.3
	SUM	32.5	55.4

face changes were detected when mining was 64 m away from the TDR hole. No surface movement was measured at that time.

Multiplexing—Strata Movement Over an Abandoned Mine

In the case of important structures located over abandoned mines, it is not adequate to say that subsidence may occur. A means must be provided to indicate if strata movements are occurring (and the rate at which they are occurring) beneath these structures so that appropriate measures can be taken to mitigate damage. The effectiveness of early warning technology inherently requires that the timing and extent of a subsidence event can be anticipated. In this case, TDR instrumentation is being employed to supply the necessary data on a continuous basis.

Coal mining was active in the Collinsville, Illinois area from 1870 to 1964, and the area is underlain by a network of mine open-

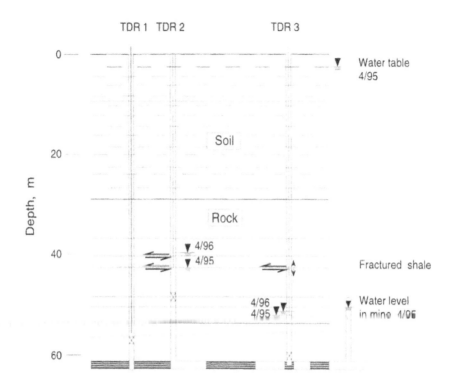

Figure 6-9. Subsurface conditions and TDR monitoring cable locations above an abandoned coal mine. Water levels in TDR 2 and TDR 3 are the maximum height of water penetration within the coaxial cables (from O'Connor, 1995).

ings. Support for the overlying rock is provided by remnant pillars and blocks of coal. When pillars have failed by crushing or by punching into the underlying claystone, or roof failure has occurred in mine entries, there have been localized mine subsidence occurrences in the city. Movement of the overlying rock, and ultimately the surface, has subjected structures, streets, and utilities to strains and stresses that have caused damage. The movements have occurred in what has been regarded as a random manner.

Hole locations shown in Figure 6-9 were selected to maximize the value of data received and the likelihood of detecting subsurface movements before they show up at the surface. The locations were selected based on the following criteria: 1) historical subsidence data, 2) proposed location for a new school, 3) subsidence currently developing in this area, and 4) mine geometry. It was desired to penetrate mine entries in the area of high extraction ratio (TDR3) as well as in the area proposed for a new building (TDR1 and TDR2).

The mines are approximately 60 m below the surface, overlain by 30 m of glacial material and 30 m of Pennsylvanian Age rock as shown in Figure 6-9. The glacial material consists of 10 m of loess overlying silty clay till and sandy stream deposits. The topmost rock stratum is a claystone that has altered to a silty clay of high plasticity. This altered zone is approximately 8 m thick and washed out easily during drilling. Clay and shales extend down to two stiffer limestone strata. The upper one is about 1.2 m thick, and the lower one is 6 to 7 m thick. This lower unit forms the immediate mine roof. No fractures were observed in core samples obtained in these limestone strata.

After drilling, sampling, and camera inspection was completed in each hole, a plug was placed to seal the bottom. A 22.2 mm diameter CommScope cable and 25 mm diameter grout tube were lowered into the hole. Prior to installation, the cable was crimped at 15 m intervals to provide reference TDR reflections. The grout tube was placed in the hole after the cable and a grout mix consisting of Portland cement, water, and an expansion agent was tremied through the grout tube. Then the grout tube was removed. The cables do not extend to mine level due to problems encountered during installation. In particular, the weathered claystone continually squeezed and plugged the holes. In hole TDR3, it was necessary to install the cable and grout tube through the wireline casing.

Lead wires were connected to each of the three coaxial cables and brought to a central location where a utility pole was installed. A TDR data acquisition system was installed within enclosures mounted on this pole. The lead wires were connected to a multiplexer. The multiplexer was connected to a TDR cable tester and a datalogger, which was connected to a storage module and modem. The datalogger was programmed to turn on the cable

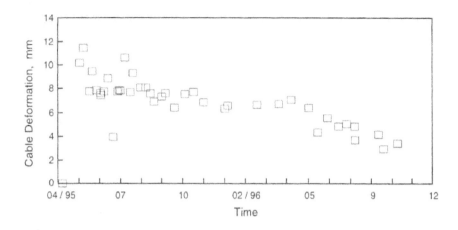

Figure 6-10. Time history of cable deformation at a depth of 42 m; decrease in magnitude indicates cable is being subjected to extension (from O'Connor, 1995).

tester, interrogate each cable, store data in the storage module, and then turn off the cable tester. Data was downloaded from the storage module via phone line.

Along cable TDR3, a reflection has developed at a depth of 42 m. This spike is characteristic of cable shear deformation. The reflection magnitude was converted to shear displacement using the calibration in Figure 5-13, and the time history is plotted in Figure 6-10. A displacement of 8 mm occurred between April and May of 1995. There was no further movement until April of 1996 when the reflection began to decrease in magnitude. This is indicative of a transition from shear to extension.

This deformation is occurring at the location of a fractured shale which is less stiff than adjacent strata. This observance is consistent with findings in studies conducted over several high extraction coal mines in Illinois (Siekmeier and O'Connor, 1994). These studies, as well as those conducted by others (Bauer et al., 1991), have found that cable deformation occurs along discontinuities between strata which differ greatly in stiffness. The low stiffness of the shale at a depth of 42 m is primarily due to the existence of fractures.

Production Well Deformation—*In Situ* Sulphur Mine

TDR was used to monitor rock mass movements induced by the Frasch hot water process during mining of strata-bound sulfur (O'Connor, 1989). A Frasch production well is drilled with a conventional oil-well drilling rig, and large quantities of super-heated water are pumped into the sulfur-bearing formation. This liquefies the sulfur, which is then lifted to the surface with compressed air. A production well consists of a series of four (or five) concentric pipes to pump hot water into the formation and return molten sulfur to the surface. As a consequence of removing the sulphur and weakening the ore-bearing strata, subsidence occurs and production wells are often sheared off in advance of the active mining front.

The mining company gained understanding of this problem by instrumenting an existing well with a cable and monitoring with TDR. As shown in Figure 6-11, a 22.2-mm diameter jacketed CommScope cable was attached to a 73-mm diameter drill pipe and lowered downhole. The cable and pipe were then grouted into the 244 mm diameter casing.

During the period from June 1989 to May 1990, a TDR reflection developed at a depth of 68 m indicating that shear deformation of the casing and cable was occurring where a breccia zone is in contact with a stiff underlying limestone (Figure 6-11). The time history is plotted in Figure 6-12 where the reflection magnitude (left axis) has been converted to shear displacement (right axis) using the calibration in Figure 5-5c (since it was the only one available at that time). Given the difference in scale between a grouted pipe in the laboratory and a grouted production well, there is bound to be some error in the conversion, but the capability of TDR to provide understanding of the subsidence mechanism responsible for well shearing was demonstrated. In cases where refinement of the conversion is desired for purposes of production monitoring, it will be necessary to set up a large-scale calibration in the laboratory.

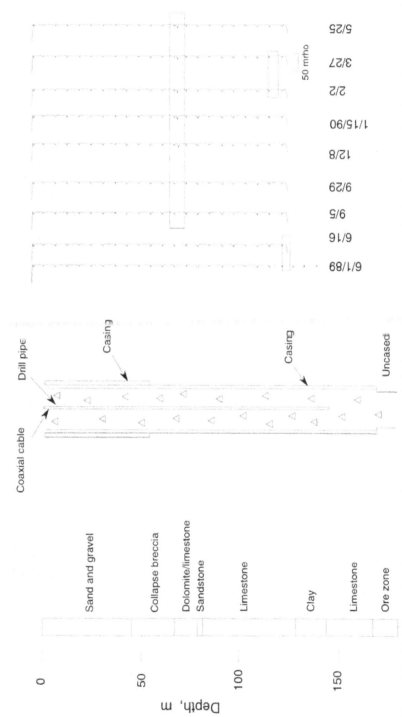

Figure 6-11. Lithology of rock mass, installation details, and TDR waveforms for cable at *in situ* sulphur mine (from O'Connor and Norland, 1995).

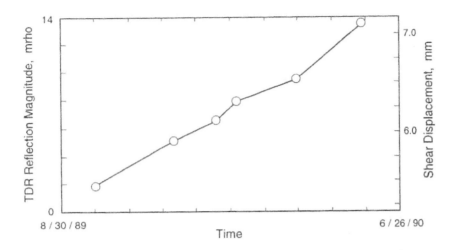

Figure 6-12. Time history of cable deformation at a depth of 68 m (from O'Connor and Norland, 1995).

Chapter 7
MONITORING
SOIL DEFORMATION

It is possible to install a compliant cable/grout system that will deform with soil shear zones in a manner similar to that observed with shearing along rock joints. Compliant systems as well as direct transfer of localized soil shearing to the cable allow automation of the monitoring process for detection of thin shear zones that accompany slope instability in a wide variety of weathered and soft rock as well as soil. In addition to developing a rationale for the use of compliant cable, this chapter presents several cases that demonstrate use of TDR monitoring of cable deformation in these softer materials.

Remote monitoring of slope movement in soils has been inhibited by the nature of manually operated instruments commonly employed for this task, namely, the inclinometer or slope indicator. Inclinometers include down-hole electronics that must be lowered down a borehole and are thus difficult to operate remotely. TDR monitoring of cable deformation can, by its digital nature, be accomplished automatically and remotely, in addition to allowing detection of very thin shear zones.

LOCALIZED SHEARING DEFORMATION OF SOIL

Localization of strains and/or narrow shear bands has been observed in the field beneath embankments, beside braced excavations, and within unstable slopes in a variety of soils including soft, saturated clays. Under large stress gradients in these environments, soil strains become significantly large and concentrated within a thin shear zone, or shear band, resulting in progressive failure of the soil. For instance, Bjerrum (1967) first described the mechanism of progressive failure of overconsolidated, plastic clays under drained conditions. In his study, slope failures were prefaced by development of continuous sliding surfaces arising from the progressive

failure mode.   Strain softening of the clay and non-uniform
stress–strain response of the soil mass were shown to give rise to
progressive failure.   Burland, Longworth, and Moore (1977)
studied the behavior surrounding a 29 m deep excavation in
overconsolidated Oxford clay with precise survey techniques and
instrumentation, including horizontal and vertical extensometers,
inclinometers, and piezometers.   Measurements indicated that the
ground behind the clay face appeared to move as a block into the
excavation, sliding along a shear band formed by bedding planes
near the base of the clay pit (Burland, 1973).   As the excavation
was advanced, inclinometers became kinked at progressively
decreasing depths away from the face.   When inclinometer casings
were removed, they showed that 20 to 30 mm of shear deformation
developed in narrow zones, indicating that intense shearing
occurred in these localized zones.

In another study, Finno et al. (1989a,b) observed a shear
band while recording ground movements adjacent to a 12 m deep
braced excavation in saturated, soft-to-medium Chicago clay.
Large but relatively homogeneous soil movements and ground
surface settlements on the instrumented side of the excavation
indicated that a distinct shear band formed within the clay mass as
shown in Figure 7-1. Movements along the shear band produced a
tension crack at the surface. The observed incremental displacements are indicative of block-type movement that is associated with
development of shearing surface and are similar to movements
observed by Burland et al. (1977). Location of a shear zone was
estimated by calculating soil strains from measured inclinometer
deformations. Strain magnitudes were calculated from displacements at 60 cm intervals along the vertical inclinometer casing. The
highest strains were concentrated along the dashed line shown in
Figure 7-1.

ADVANTAGES OF TDR MEASUREMENT OF LOCALIZED SHEARING IN SOIL

Inclinometers are normally employed to measure shearing deformation in soil. This measurement is made normal to the axis of a
vertical casing with a gravity sensing probe as shown in the

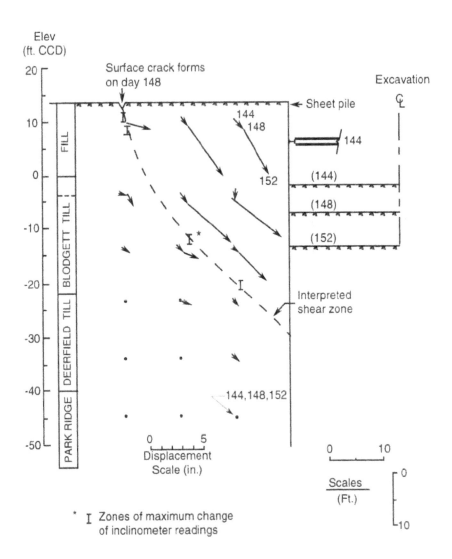

Figure 7-1. Approximate shear zone location adjacent to a braced excavation in soft, saturated clay. Numbers indicate days since excavation began (from Finno et al., 1989a).

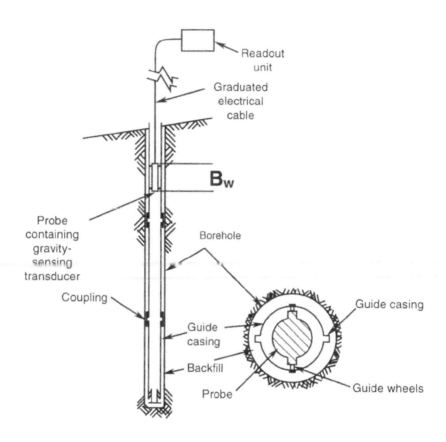

Figure 7-2. Cross-sectional view of inclinometer and probe (from Dunnicliff, 1988; reprinted by permission of John Wiley & Sons, Inc.).

detailed sketch in Figure 7-2 (Dunnicliff, 1988). As the wheeled probe travels within grooves along the casing, a servo-accelerometer measures inclination of the probe. By cumulatively summing the product of the inclination and the probe length as the probe is pulled up from an assumed fixed position, the casing profile can be determined.

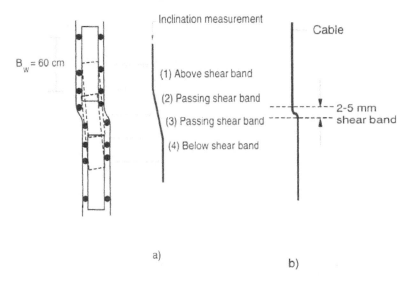

Figure 7-3. Resolution of shear band measurements; (a) using an inclinometer, (b) using TDR (from Dowding and Pierce, 1994a).

Probe movement shown in Figure 7-3a demonstrates that the resolution of shear band detection with an inclinometer is controlled by the wheelbase, B_w, which is typically 60 cm. In this example, it is assumed that the casing deforms conformably with a thin shear band approximately 5 mm thick. As the probe is incrementally lowered, it can measure the angle across the shear band (four increments are indicated by the dotted positions of the probe). Resolution of band thickness is limited to the wheelbase length, even if readings are taken at intervals that are a fraction of this length (McKenna, 1995). Furthermore, thin shear bands can be misinterpreted, for example, if the probe wheels do not move smoothly along the grooves. Aside from the low sensitivity to localized shear due to probe geometry, several other disadvantages arise when using an inclinometer to detect thin zones of deformation. One concern is that the casing may be too stiff to conform with soil deformation along thin, transverse shear bands in soft soils. Also, another limitation occurs when deformation at a shallow but distinct shear zone prevents passage of the probe to deeper sections and precludes measurement altogether.

Use of an appropriately compliant cable/grout system should allow measurement of shear zones as thin as 5 mm, as shown in Figure 7-3b. This detectable thickness is approximately (5 mm/ 600 mm =) 1/120 that possible with an inclinometer. Depending upon the cable length and TDR rise time (discussed in Chapters 2 and 4), it is possible to easily detect such localized deformation. TDR techniques can be effectively used to detect cable shear events separated by distances as small as 200 mm at cable lengths up to 94 m (Table 5.3 and Figure 5-14). Single shearing events along a cable can be detected hundreds of meters from a TDR pulser; however, the ability to distinguish or resolve closely spaced shear incidences beyond 100 m requires further study (Figure 5-10).

TDR has several other advantages over inclinometers. The most important advantage is complete automation of data acquisition, as shown by its successful use to remotely monitor rock deformation over active and abandoned coal mines as discussed in Chapter 6. Multiplexing allows multiple cables to be monitored from electronics installed at a central location. These advantages show that compliant cable grout systems may be deployed for remote, early detection of subsurface movements in any number of situations using TDR.

REQUIRED GROUT AND CABLE COMPLIANCE

Different failure modes must be considered to assess soil–grout–cable interaction. Either localized (Figure 7-4a) or nonlocalized failure may occur within the soil. A nonlocalized failure mode indicates generalized soil shearing that produces a relatively wide shear band in the deforming mass, as shown in Figures 7-4b and 7-4c. Such deformation may fracture the grout backfill around a cable, but may not be sufficiently concentrated to locally shear the cable and would not produce a detectable TDR reflection. However, past observations of such nonlocalized deformation could actually represent groups of thinner, localized shear bands. Use of TDR should allow such a distinction between localized and nonlocalized failure.

As discussed earlier, localized failure results from non-uniform, concentrated strain along a shear band in the soil. When

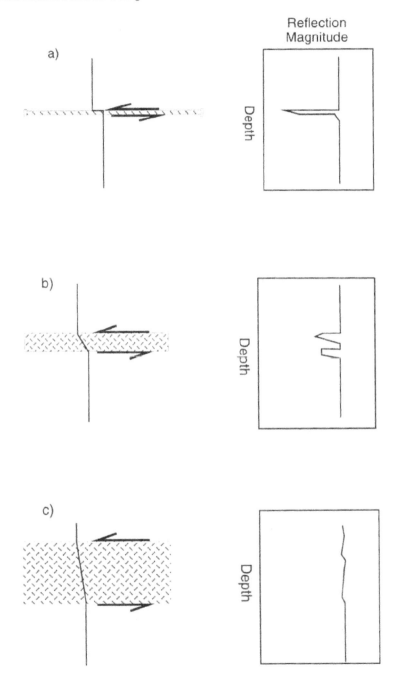

Reflection
Magnitude

Figure 7-4. Characteristic TDR waveforms for various shear zone widths: (a) localized shear; (b) shear zone with cable bending; (c) large shear zone (from O'Connor et al., 1995).

such a shear band intersects the cable/grout column and fractures the grout, it then deforms the cable. As discussed in Chapter 5, the amplitude of the TDR reflection will be proportional to the magnitude of local cable deformation. The cable will deform and ultimately be severed, thereby producing a classic end-of-cable spike reflection in the TDR waveform as shown in Figure 5-8.

In addition to the shear zone width, yielding and flow of the soil around the grouted cable is another consideration. In the case of a relatively strong and stiff cable/grout system, the soil movement may not deform the cable sufficiently to cause a detectable TDR reflection. For purposes of discussion, assume that the strength and stiffness of a cable/grout system can be designed to equal the strength and stiffness of the surrounding soil. In order for deformation to be detected using TDR, the grout must fracture and start deforming the cable as the soil yields plastically within the shear zone.

To accurately detect soil deformation along a localized shear surface, the cable must deform easily when subjected to localized soil shearing. Thus it is important to consider the deformation properties of soil, grout, and cable to determine if a cable/grout system will be sufficiently compliant. Among the commercially available coaxial cables that have been used to monitor rock and soil deformation are those manufactured with solid polyethylene as the dielectric material between solid aluminum conductors. Polyethylene has a stiffness and shear strength of approximately 3×10^6 kPa (4.4×10^5 psi) and 870 kPa (125 psi), respectively (Society of the Plastics Industry, 1960). This cable works well for detection of rock movement where the intact rock is stiffer and stronger than polyethylene. A more compliant material, such as Styrofoam or polystyrene foam, with a stiffness and shear strength of 280 kPa (41 psi) and 4 kPa (0.6 psi) would be more deformable. The outer conductor also contributes to cable stiffness. While cable with a solid aluminum outer conductor is preferred for monitoring deformation in rock, an outer braided copper conductor so dramatically reduces stiffness that it may be preferable for monitoring in soil (although the TDR waveform quality may be degraded).

Compliance of the grout backfill surrounding a cable can be designed using a variety of materials including: 1) fly ash/bentonite mixtures, 2) foamed concrete, and 3) cement/bentonite mixtures. Fly ash mixtures have been found to be sufficiently compliant, but they vary in strength by source and may contain sufficient sand-sized particles to inhibit pumping with a drill-rig water pump. Foamed cements are also sufficiently compliant, but may require an operator sophistication not available in many low-volume situations. Cement/bentonite mixtures appear to be simple to mix and control, and can be designed to attain a sufficiently low viscosity to be placed using drill rig water pumps.

The following example illustrates one approach for determination of an appropriately compliant cable/grout system. Assume that a weak and compliant cable is available with a strength and stiffness of 4 and 280 kPa (0.6 and 41 psi), respectively. Also assume that a compliant grout can be produced with an unconfined strength of 100 kPa (14.4 psi) and a stiffness of 3×10^5 kPa (4.4 x 10^4 psi) as shown in Figure 7-5c. It is then necessary to compare these strengths and stiffnesses with those of the soil.

Localization of deformation has been observed to occur in medium to stiff overconsolidated clays (Burland et al., 1977; Finno et al., 1989a,b). A medium stiff clay, for example, may have a shear strength of 100 kPa (14.4 psi) and stiffness of 1×10^5 kPa (1.5×10^4 psi). Shear strength of the hypothetical grout described above is equal to one-half of its unconfined compressive strength or (½ x 100 kPa =) 50 kPa (7.2 psi). Grout shear strength would then be less than the shear strength of medium clay and the grout stiffness would be only slightly greater than that of the clay. Hence the hypothetical cable/grout system would be expected to fail and deform conformably during localized shearing in such a medium strength clay.

Even soft clay could theoretically fracture grout that has a shear strength of 70 kPa (10 psi) before flowing in general plastic failure. Finno et al. (1989a,b) observed failure along a localized shear band which occurred in soft clay (Blodgett Till) with an undrained strength at 24 kPa (3 psi) and stiffness of 1.2×10^4 kPa (1.7×10^3 psi) as shown in Table 7.1b. To ensure that the grout fails first, general plastic bearing failure of the soft clay around the cable/grout cylinder, as shown in Figure 7-5b, must not occur. The general bearing capacity equation (Terzaghi and Peck, 1967) is

$$q = c\,N_c + \gamma_{bulk}\,D_f\,N_q + 1/2\,\gamma_{bulk}\,D\,N_\gamma \qquad (7\text{-}1)$$

where N_c is the shear strength factor (= 9), N_q is the embedment depth factor, N_γ is the bearing area shape factor, γ_{bulk} is the bulk unit weight of soil in units of N/m^3, D_f is the embedment depth in units of m (Figure 7-5a), and D is the borehole diameter in units of m (Figure 7-5b). For soft clay, the contributions of embedment and bearing area size are ignored. To ensure that grout failure occurs within the elastic range of the clay as shown in Figure 7-5c, cable/grout contact pressures should not exceed 1/3 of the clay bearing capacity, 9c, where c is ½ of q_u, unconfined compressive strength (Peck et al., 1974). Therefore, if the grout shear strength is less than (9c)/3 = 3c of the soil, the grout will fail first. For the Blodgett Till, 3c is approximately 72 kPa (10 psi).

While bearing capacity provides one approach to grout-cable design, it does not adequately address the issue of relative stiffness or compliance. The approach of soil–structure interaction is addressed later in this chapter after considering experience gained from monitoring of slope movement.

LOCALIZED VERSUS GENERAL SHEAR

When cables are grouted into precut pipes and subjected to direct shear as discussed in Chapter 5, there is very localized fracturing of the grout, and cable deformation occurs within a shear zone less than 5 mm thick, as shown in Figure 7-4a. It is this type of localized deformation for which TDR sensitivity will be greatest. A second mode of deformation that can be detected is bending of the cable, if it is bent through a radius of curvature smaller than a critical value, as shown in Figure 7-4b. Large shear zones can be detected, provided such bending occurs. Bending of cable is discussed in connection with field applications and also in connection with the use of gravel backfill. The TDR sensitivity to cable bending is much less than sensitivity to localized shearing.

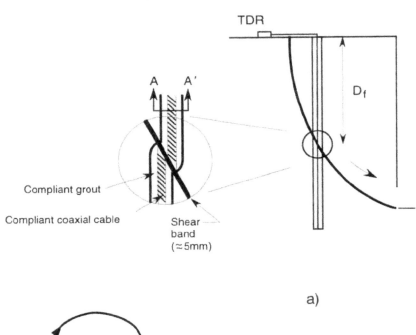

a)

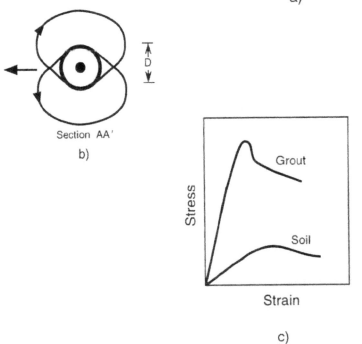

Figure 7-5. Considerations for designing a compliant cable/grout system:
(a) installation; (b) generalized bearing failure within the soil; (c) relative
stiffness and strength.

Table 7.1a. Comparison of strength and stiffness of grouted cable.

Cable or Material	Backfill	Diameter	Stress at Yield (kPa)	Stress at Yield (psi)	Strain at Yield	Stiffness (MPa)	Reference
9.5 mm FXA 38-50	cement	51 mm precut steel pipe mold	1010	146	0.0343	29	Su (1987)
12.7 mm FXA 12-50			1010	146	0.0298	29	
22.2 mm FXA 78-50			2800	403	0.1687	16	
12.7 mm FXA 12-50		76 mm PVC pipe mold	4600	665	0.0066	697	Peterson (1993)
19.1 mm P3 75-750CA		89 mm clamped 40 mm shear zone	1590	230	0.0180	88	
12.7 mm FXA 12-50	crushed gravel	200 mm steel pipe	600 to 900	87 to 130	0.1250	5 to 130	Logan (1989)

Table 7.1b. Comparison of strength and stiffness
of dielectric materials and clay.

Cable or Material	Stress at Yield		Strain at Yield	Stiffness	Reference
	(kPa)	(psi)		(MPa)	
Polyethylene	870	126	0.0003	3000	Society of Plastics Industry (1960)
Polystyrene	4	0.6	0.0143	0.3	
Blodgett Till	24	3	0.0020	12	Finno et al. (1989a,b)

APPLICATIONS

The potential advantages of TDR for use in soil has motivated a few comparative studies in which it was used in conjunction with inclinometers to monitor subsurface movements in unstable slopes. One study was conducted at a Canadian oil sands mine to monitor movements in a highwall slope and in a tailings embankment. Other studies have been conducted in California to monitor movements in active landslides.

Highwall Slope and Tailings Embankment

Syncrude Canada, Ltd. operates an open-pit oil sands mine near Ft. McMurray, Alberta that requires an extensive slope monitoring program and investigated the use of TDR. The bituminous sand is excavated by four large (70 m³) draglines which sit near the highwall and must be protected against undermining by a highwall failure. The bituminous sands contain numerous thin, consolidated clay layers that can cause highwall slope instability. The highest-risk highwall-failure mode is a block slide, which can occur along any estuarine clay layer that dips into the pit at an angle greater than

10° from horizontal. A block slide can occur after only a 10 mm to 15 mm displacement along a clay layer and moves at a critical velocity of about 1 mm per hour just prior to failure. Other failure modes include: a) slides along flat-lying marine clay layers due to loading of the highwall with ore stockpiles, and b) flow failures of rich oil sand, which cause deep tension cracks and undermining of the dragline bench. Since mining started in the late 1970s, there have been about 40 block slides, extensive flow failures, and a variety of other slope failures.

A slope inclinometer system is used extensively in the mine and tailings areas to monitor ground movement. The position and spacing of inclinometers is determined by the Highwall Engineer, after considering the highwall geology and the experience gained while mining the area previously (Fair and Lord, 1984; Holmes and List, 1989). The number of times that inclinometers are read is increased when critical movement (2.5 mm) is detected, and an increase in the rate of movement to 0.5 mm/hr will justify continuous monitoring of a specific movement zone. Slope inclinometers have inherent safety and operational concerns: 1) dragline operations are stopped when reading inclinometers within the dragline boom radius (25 m from the dragline tub), 2) Highwall Monitoring Engineers are placed in a less safe situation when the inclinometer holes are located between the dragline and the potentially unstable highwall crest, and 3) in an active slope movement situation, even increasing the number of readings does not ensure adequate notice of an impending highwall failure.

As summarized in Table 7.2, coaxial cables for TDR monitoring were installed at three highwall locations and at two locations in the tailings embankment. The installation in each case was in the immediate vicinity (less than 3 m) of a slope inclinometer so that a comparison could be made between the two types of instrumentation. The cable installed in a horizontal hole drilled into the tailings embankment was in the immediate vicinity of a horizontal extensometer. The objectives of these field tests were to assess ease of installation, suitability to field conditions, ease of data acquisition, a comparison with existing monitoring procedures, and sensitivity of TDR to slope movements.

The highwall installation (cable M01) shown in Figure 7-6

Table 7.2. TDR Cables installed at Syncrude Oil Sands Mine (from Peterson, 1993).

Hole	Cable	Length down hole (m)	Lead length (m)	Location	Failure mode
T01	22.2 mm Cablewave FXA 78-50	54.9	0	vertically into tailings embankment	shear along clay zone beneath embankment
M01	12.7 mm Cablewave FXA 12-50	24.3	120	highwall	block failure
M02	19.1 mm CommScope P3 75-750CA	22.9	107	highwall	block failure
M03	same	49.5	0	highwall	bulging failure
H01	12.7 mm CommScope P3 75-500CA	50	0	horizontally into tailings embankment	

All cables were installed in 143 mm diameter holes.

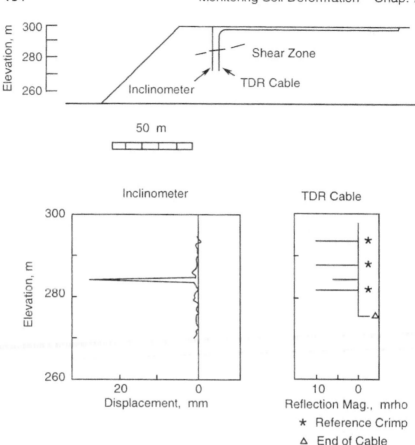

Figure 7-6. Location of coaxial cable and inclinometer in highwall slope and indication of movement at depth of 18 m (from O'Connor et al., 1995).

was made in a 140 mm diameter hole drilled to a depth of approximately 26 m. A 1 m deep backhoe trench was dug extending 120 m from the top of the hole. A 12.7 mm diameter (Cablewave FXA 12-50) coaxial cable was crimped at 3 m intervals, run through PVC conduit, laid out in the trench, and lowered into the hole. The three crimps produced TDR reflections which appear at elevations of 294, 288, and 282 m in the lower right side of Figure 7-6. The cable was grouted into place by pumping a standardized grout mix (11 bags of cement [25 kg each], 162 liters of water, and 1 bag of bentonite [19.5 kg each]) through a tremie pipe. After grouting the hole, the trench was backfilled so that monitoring could continue as the dragline moved past this location. The total cable length was

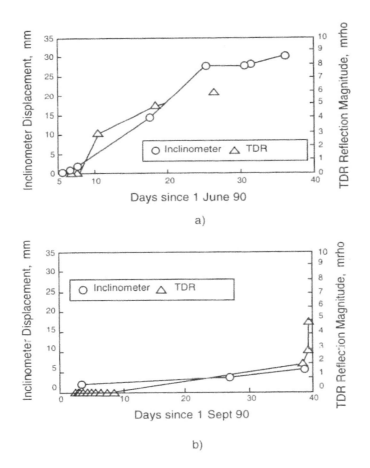

Figure 7-7. Time histories of inclinometer displacement and TDR reflection magnitude (from Peterson, 1993).

145 m.

In agreement with the inclinometer, a TDR reflection developed at an elevation of 283 m. Both the inclinometer displacement and TDR reflection magnitude increased as slope movement occurred over a period of 40 days as shown in Figure 7-7a and Table 7.3. Laboratory calibration tests in which the cable was deformed within a shear bend less than 5 mm thick indicated that the TDR sensitivity was typically 20 mρ/mm as shown in Table 5.1. However, when the TDR reflection magnitude in Table 7.4 is

Table 7.3. Comparison of TDR and Inclinometer (from Peterson, 1993).

Hole	Deformation	Inclinometer response	TDR response
T01	movement beneath embankment	several mm over 2 years along clay layer at depth of 38.1 m	no change over 2 years
M01	block failure	shear along clay at depth of 16 m	shear along clay at depth of 16 m
M02	block failure	shear at depth of 17 m	shear at depth of 15 m; water penetrating cable
M03	bulging failure	detected movement over zone extending from 12 to 17 m, then no movement in this zone; ultimately detected shear at depth of 33 m	detected shear at depth of 33.4 m

Table 7.4. Compilation of inclinometer displacement and TDR reflection data.

Inclinometer displacement (mm)	TDR reflection magnitude (mrho)	TDR sensitivity (mrho/mm)	Distance from TDR (m)	Strata type	Cable type	Reference
18.00	18	1.0	49.7	rock:	A	Kawamura et al. (1994)
3.00	5	1.7	59.7	coal measures		
8.00	18	2.3	61.9			
23.00	8	0.3	67.7			
28.00	7	0.3	77.4			
14.00	9	0.6	89.0			
0.82	0	-	136	soil:	A	Peterson (1993)
0.99	0	-		interbedded		
1.01	0	-		sand/silt/clay		
1.68	0	-				
14.31	5	0.3				
27.51	6	0.2				

Cables:

A = Cablewave FXA12-50, smooth aluminum outer conductor, foam polyethylene dielectric.

B = CommScope P3 75-750CA, smooth aluminum outer conductor, expanded polyethylene dielectric.

C = Belden RG59/U, braided copper outer conductor, Teflon dielectric.

Table 7.4. (continued)

Inclinometer displacement (mm)	TDR reflection magnitude (mrho)	TDR sensitivity (mrho/mm)	Distance from TDR (m)	Strata type	Cable type	Reference
2.38	0	-	122	soil:	B	Peterson (1993)
1.88	0	-		interbedded		
5.40	2	0.4		sand/silt/clay		
5.92	5	0.8				
5.92	3	0.5				
25.0	16	0.6	67	soil:	C	Kane (1998a)
241.3	48	0.2		soft peat		

Cables:

A = Cablewave FXA12-50, smooth aluminum outer conductor, foam polyethylene dielectric.
B = CommScope P3 75-750CA, smooth aluminum outer conductor, expanded polyethylene dielectric.
C = Belden RG59/U, braided copper outer conductor, Teflon dielectric.

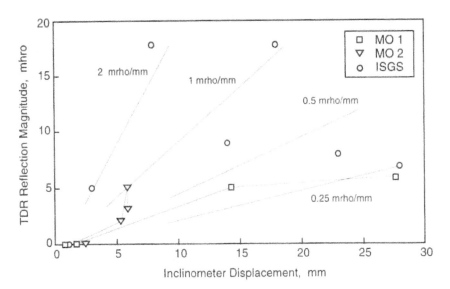

Figure 7-8. TDR field data obtained with solid aluminum coaxial cable and corresponding inclinometer displacement from Table 7.4.

plotted versus the inclinometer displacement, it appears that the field sensitivity was only 0.2 to 0.5 mρ/mm as shown in Figure 7-8. There are several factors which can contribute to this decrease in sensitivity, including a) distance along the cable from the TDR pulser as discussed in Chapter 5, b) shear zone width, and c) deformation mode as discussed below.

Another installation (cable M02) was similar to cable M01 with the cable grouted into a 143 mm diameter hole and buried in a trench approximately 107 m long. A 19.1 mm diameter CommScope cable was used and it was crimped at three locations. While the inclinometer showed a movement zone at a depth of 17 m, the cable indicated movement at a depth of only 15 m (Table 7.3). For several reasons, the depths were not referenced to the crimps made in the cable, so this discrepancy cannot be reconciled. However, the time histories of inclinometer displacement and TDR reflection magnitude are consistent as shown in Figure 7-7b. When the TDR magnitude is plotted versus inclinometer displacement as shown in Figure 7-8, the field sensitivity is again in the range of 0.2 to 0.5 mρ/mm.

The much lower field sensitivity of TDR compared with lab-

Table 7.5. Effect of shear zone width on sensitivity and cable[1] deformation (Peterson, 1993).

Shear Zone Width (mm)	TDR Sensitivity (mrho/mm)	Mode of Cable Deformation
0	19.66	localized (Figure 7.4a)
5	14.86	
20	8.28	
40	9.91	
80	0.96	bending (Figure 7.4b)

[1] FXA12-50, 12.7 mm, solid aluminum, Cablewave Systems, North Haven, CT.

oratory sensitivity is evident by comparing sensitivities of the compiled data in Figure 7-8 with the sensitivities in Table 5.1. A series of tests were performed in which the shear zone width could be adjusted. The results are listed in Table 5.1, Table 7.4, and Table 7.5. Based on the results of tests by Logan (1989) with crushed gravel backfill in simulated boreholes which are discussed later, and a postmortem investigation of Syncrude grouted cable lab test specimens, it was also possible to identify the mode of deformation. As the mode changed from localized shear (Figure 7-4a) to bending at the top of the shear zone (Figure 7-4b), the characteristics of the TDR signature also changed as shown on the right side of the figure.

Relative Stiffness Revisited

The question then becomes "What were the relative strengths and stiffnesses?" Data compiled in Tables 7.1 and 7.6 indicate that grout strength and stiffness were in the elastic range of 1000–2000 kPa and 50–150 MPa. If the clay seams in the oil sands were similar to soft Blodgett Till, then their strength and stiffness would be 24 kPa and 12 MPa. Since the grout was significantly stiffer (by an order of magnitude), there would have been movement of the grout column relative to the clay mass on either side of the shear

Table 7.6. Comparison of strength and stiffness of grout and soils (Fair and Lord, 1984; List, 1992; McKenna, 1997).

	Failure Mode	Materials	Shear Strength	Stiffness
19.1 mm cable P3 75-750CA	n/a	SLC grout	1590 kPa	88 MPa
tailings	movement along clay seam beneath embankment	clay seam	n/a	n/a
highwall block	failure along clay seams	interbedded sand/silt/clay	$c = 0$ to 18 kPa $\phi_{peak} = 13°$ to $29°$ $\phi_{res} = 8°$ to $25°$	40 to 125 MPa*
highwall bulge	general shear due to reduction in confinement	interbedded sand/silt/clay	$c = 0$ to 18 kPa $\phi_{peak} = 13°$ to $29°$ $\phi_{res} = 8°$ to $25°$	40 to 125 MPa*

*Assumptions: $\sigma = 200$ kPa; $\tau = c + \sigma \tan \phi$; $E/\tau = 1000$.

surface prior to fracture of the grout. Note the bearing capacity of the soft clay would only have been $3c = 3 \times 24$ kPa $= 72$ kPa, significantly less than the strength of the grout-encapsulated cable. Similarly, medium strength clay would have had a bearing capacity of only 3×100 kPa $= 300$ kPa. Thus, even for a medium strength clay, general bearing capacity failure would have occurred within the clay as shown in Figure 7-5b and relative deformation would occur before grout fracture.

This experience shows the importance of relative strength and stiffness of the clay/grout/cable system as simplistically shown in Figure 7-5c. It is hypothesized that the grout column behaves as a higher-strength, stiff inclusion which rotates relative to the soil as shown in Figure 7-9b. Such rotation spreads the relative deformation along the grouted cable column.

Analogous to this situation is the shear resistance and behavior of soil nails. Juran (1986) reports an approach to passive resistance of a rigid inclusion for soil nails or laterally loaded piles, as shown in Figure 7-9b. This approach is based on the problem of an infinite elastic beam subjected to bending in a Winkler soil, where the beam's reaction is governed by the soil's coefficient of subgrade reaction.   With some simplifying assumptions, this approach can be adopted for determining the resistance of any stiff inclusion such as a grout-encapsulated cable. The shear force V_s acting on the grouted cable is limited by what the soil can mobilize (Elias and Juran, 1991; Xanthakos et al., 1994),

$$V_s = 1/2(pDL_0) \tag{7-2}$$

where p is the passive (normal) pressure acting on the grouted cable in units of N/m^2 and D is the borehole diameter in units of m. The maximum value of p may be taken as the "creep pressure" as measured in a pressuremeter test or may be estimated using Equation (7-1). The so-called transfer length

$$L_0 = [(4EI_0)/(k_sD)]^{1/4} \tag{7-3}$$

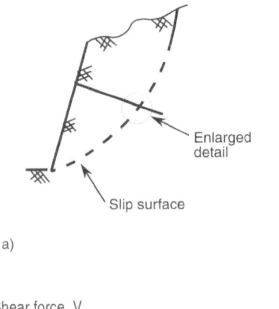

Enlarged
detail

Slip surface

a)

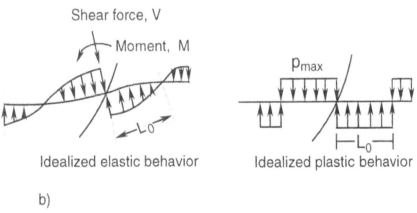

Shear force, V

Moment, M

$-L_0$

Idealized elastic behavior

p_{max}

$-L_0-$

Idealized plastic behavior

b)

Figure 7-9. Soil-structure interaction model for a soil/grout/cable system: (a) slip surface along which localized shear occurs; (b) sketches illustrating development of passive resistance of soil (from Juran, 1986).

is shown in Figure 7-9b, where E is the grout stiffness in units of N/m^2, I_o is the moment of inertia of the grout/cable cross section in units of m^4, and k_s is the modulus of subgrade reaction of the soil in units of $N/m/m^2$.   Studies have been conducted (Dowding and Pierce, 1994a) to evaluate the applicability of such a soil-structure interaction model, and these are continuing.

Landslide Monitoring

The California Department of Transportation (CALTRANS) has placed several coaxial cables alongside inclinometer casings at several sites as summarized in Table 7.7 (Kane and Beck, 1996b). To identify the depth of the slide plane before recommending a repair/rerouting strategy at the Last Chance Grade location, Caltrans installed two inclinometers in the winter of 1994. One inclinometer was installed in the southbound traffic lane. Attached to the outside of the inclinometer casing was a length of 12 mm diameter braided outer conductor cable for TDR monitoring. A groove was cut in the pavement and the cable extended off the shoulder of the roadway to a location behind the adjacent concrete guard rail in a manner similar to that shown in Figure 4-8.

The advantage of the TDR installation was that the cable could be read in a manner of minutes by a single technician safely behind the barrier without any interruption in traffic flow. By contrast, reading the inclinometer required a crew of three to four people to reroute traffic to a single lane on a very heavily traveled roadway. The asphaltic concrete pavement (ACP) then was removed from above the casing, the inclinometer read, and the ACP patch replaced. This process required several hours.

A second location, the Grapevine Slide, is in a steep and remote area and involved a remote data acquisition and communication system. To read the inclinometers, two technicians traveled two hours in one direction from Fresno, then climbed a difficult and dangerous slope. Since each inclinometer was approximately 90 m deep, requiring several hours to read, the entire polling process took a full day. The remote data acquisition system installed to monitor the cable included a cable tester, data logger, multiplexer, solar panel, cellular phone, and modem. When fully operational, the system enables movement to be monitored via phone from any computer.

At a third location, Devil's Slide, the rapid slope movements could expose the polling technician to landslide or rockfall danger. This danger prevents real-time monitoring of the slide during periods of most interest to the geotechnical engineer. Deploying TDR cables in such a situation allows data collection

Table 7.7. CALTRANS Landslide Monitoring (Kane and Beck, 1996b; Kane et al., 1996; Freeman, 1996).

	Location	Installation	Response
Last Chance Grade Landslide	U.S. Hwy. 1, Del Norte County	Type A cable attached to inclinometer casing (backfilled with sand); hole 80 m deep	Displacement at a depth of 40 m detected by both inclinometer and cable
Grapevine Grade Landslide	U.S. I-5, Kern County	Type A cables attached to inclinometer casing; one cable attached to piezometer; remote monitoring system	Cables did not indicate any movement
Cloverdale Landslide	U.S. Hwy. 101, Sonoma County	Type A cables attached to two of four inclinometer casings	Neither inclinometers nor cables indicated movement
Willets Landslide	State Hwy. 20, Mendocino County	Type B cables attached to inclinometer casings	Cable could still be monitored after one of the inclinometers was pinched off
Devil's Slide	U.S. Hwy. 1, San Mateo County	Type A and Type C cables attached to one inclinometer	

Cables:
Type A   7.6 mm diameter, RG-6/U braided copper, Radio Shack.
Type B   6.2 mm diameter, RG-59/U braided copper, Belden.
Type C   12 mm diameter, FLC 12-50J corrugated copper, Cablewave Systems.

safely away from the slide. It also is possible that automatic remote polling, processing, and communication of TDR signatures can serve as an early warning system. In other words, TDR systems can help sense impending catastrophic movement rather than reporting after-the-fact movement, as with the present warning system.

CALTRANS experience with TDR has shown that it is a cost-effective alternative to inclinometers in many instances (Kane and Beck, 1996b). It provides quick and easy monitoring of slope movement, can accurately locate shear planes, and provides an indication of changes in the rate of movement. In the future, TDR will be installed to provide early warning systems for movement. TDR cables also have the potential to be installed in small diameter boreholes or casings that have deformed too much for inserting an inclinometer probe.

Kane (1998b) reports that a further innovation has been the use of inexpensive electrolytic tilt sensors which are grouted into the boreholes along with the coaxial cable. These tilt sensors increase the sensitivity to movement in slopes and excavations experiencing general failure, such as that which occurred at location M03 in Table 7.3. These tilt sensors make it possible to obtain an early warning of movement, which is then confirmed by interrogating the cables using TDR.

COMPARATIVE PERFORMANCE AND COSTS

The Syncrude experience proved that coaxial cable can be successfully calibrated, installed, and field-repaired in the highwall and tailings embankment using existing equipment, personnel, and procedures. The level of effort, cost, and time required to install a vertical coaxial cable was comparable to the level of effort, cost, and time required to install a slope inclinometer. Field personnel felt that the 12.7 mm Cablewave cable was the easiest to handle and to install. The 12.7 mm CommScope cable was unacceptably weak, there were kinking problems with the 19.1 mm CommScope cable (although installation procedures could be implemented to overcome this difficulty), and the 22.2 mm Cablewave cable was difficult to handle. A comparison of inclinometer and TDR performance is summarized in Table 7.8.

Table 7.8. Comparison of performance (Peterson, 1993).

	Inclinometer	TDR
Operating principle	Inclination of probe	Deformation of coaxial cable
Grout interaction	Grout can deform elastically	Grout must fracture
Minimum displacement	0.03 mm	1.0 mm[1] 5.0 mm[2]
Linear range	30 mm	10 mm[1] 20 mm[2]
Accuracy	± 2 mm per 25 m of casing	± 2 mm
Minimum shear zone thickness	2 mm	2 mm

[1]FXA12-50, 12.7 mm, solid aluminum, Cablewave Systems, North Haven, CT.
[2]P3 75-875, 19.05 mm, solid aluminum, Comm/Scope, Catawba, NC.

The comparison in Table 7.9 indicates some of the personnel time and hardware cost requirements. Cable has a significant cost advantage versus casing. Solid aluminum coaxial cables can be purchased for $2 to $6/m ($0.6 to $1/ft) as opposed to $15 to $30/m ($5 to $10/ft) for inclinometer casing. The cost of the electronic equipment is approximately the same—roughly $8000 per unit. For remote installations, dataloggers and cellular phone equipment run another $5000. The cost of drilling is the same. With TDR, there is no risk of losing an expensive probe due to localized deformation at any depth and only expendable cable is lost.

In addition to hardware cost savings, the advantages of automated polling are significant. While the inclinometer casing must be polled by moving the transducer shown in Figure 7-2 along the casing, TDR can be read by simply pushing a button to launch the pulse. Rapid reading of TDR cables means that more holes can be read in less time–an additional savings in personnel time. For example, a single cable can be read in about 5 minutes or less, as

Table 7.9. Cost comparison (1998 prices).

	Inclinometer	TDR
Downhole cost	$15.00/m; grooved casing	$6.00/m[1] $2.00/m[2]; coaxial cable
Sensor cost	$8000; probe, cable, and readout	$8000; cable tester
Data acquisi-tion	20 min	5 min
Data reduction	20 min	20 min

[1] FXA12-50, 12.7 mm, solid aluminum, Cablewave Systems, North Haven, CT.
[2] P3 75-875, 19.05 mm, solid aluminum, Comm/Scope, Catawba, NC.

opposed to ½ hour to an hour to profile an inclinometer casing. Multiple cables at a site can be run to an easily accessible central location and can be read from that one location. Most important, the digital nature of TDR waveforms allows remote monitoring.

GRAVEL PACK ALTERNATIVE

As evidenced by the discussion of compliant grout/cable systems earlier in this chapter, a variety of options can be explored, each of which may be more appropriate for a particular project. For example, laboratory studies have been performed to evaluate the performance of cohesionless, granular backfill rather than cementitious grout.

A cable shear tester (CST), shown schematically on the left side of Figure 7-10, was designed to simulate granular backfill in a 200 mm diameter borehole undergoing shear and also allow for a coaxial cable to be embedded and monitored (Logan, 1989). The CST consists of upper and lower immobile cylinders and a sliding middle cylinder to simulate a shear zone. The upper and lower cylinders are equipped with removable plugs to provide backfill confinement, and a hose attachment on the upper cylinder allows for

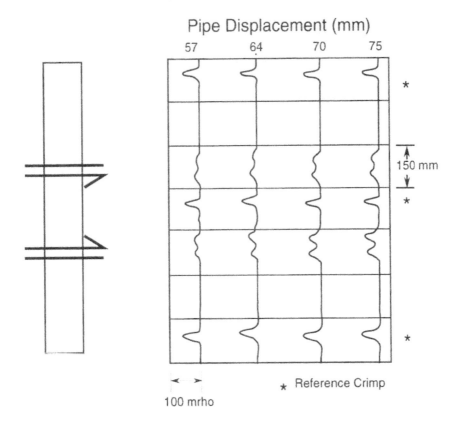

Figure 7-10. Large cable shear tester (CST) used for evaluation of gravel
backfill and TDR waveforms acquired (from Logan, 1989).

saturating the backfill during a test.   A six-threads-per-inch
displacement screw provided a simple means for controlled
displacement of the center cylinder. A torque wrench was used to
turn the screw in order to provide a measure of the shearing force
required to displace the middle cylinder.

Two backfill materials from gravel pits near Lemitar, New
Mexico were selected for the CST tests. One material was 12 mm
gravel produced by crushing of boulders, while the second material
was clean alluvial sand passing a 6 mm sieve. The sand originally
had some pebbles in it, but only the portion passing the No. 4 sieve
(4.75 mm) was used in order to have a grain size distribution
distinctly different from the crushed gravel.

**Table 7.10. Summary of test conditions for
cable in gravel backfill (Logan, 1989).**

Test Number(s)	Backfill Conditions	Number of Anchors
1 to 4	Dry Gravel	1
5 to 7	Dry Gravel	3
8	Dry Sand	1
9	Dry Sand	3
10 to 14	Saturated Gravel	1

Cable clamps were used as anchors to increase the frictional resistance between the cable and backfill. To evaluate the influence of varying backfill and cable anchor conditions, a series of tests were conducted as indicated in Table 7.10. All tests were conducted with at least one clamp attached to the cable within the bottom cylinder. For the tests conducted with 3 anchors, a clamp was attached to the cable within both the middle and upper cylinders to evaluate the resistance to cable slipping.

Typically all tests were continued until either the cable failed or the middle cylinder had displaced 100 mm. The cable was then removed from the backfill, photographed as shown in Figure 7-11, and measured to quantify cable deformation. To develop an understanding of the cable deformation process, two tests were stopped when a change in TDR reflection coefficient was first detected and the cable was removed from the backfill.

Cable deformation during a test with dry crushed gravel is shown in Figure 7-12 by considering the movement at points A, B, and C (points A and C are initially at the location of the upper and lower shear planes). As the middle cylinder was displaced to the right, a passive wedge developed within the gravel. The shape of this wedge and the position of point B remained constant. The cable was bent at two locations, at point A and also at point C as shown in Figure 7-4b, due to the moment created by the passive wedges above and below each shear plane. With continued pipe displacement, the radius of curvature at each bend decreased until

Figure 7-11. Deformed cables removed from CST test; close up of bends which developed above and below the upper shear plane.

the length of cable within each shear plane was in a state of almost pure tension. As pipe displacement continued, necking and eventual tensile failure of the outer aluminum conductor occurred. The ultimate tensile failure is shown by the cable labeled #1 in Figure 7-11. In tests conducted using sand backfill, gradual deformation of the cable occurred as shown in Figure 7-4c. The low friction angle, and resulting large moment arm between passive wedges, only caused gradual bending of the cable, which did not produce a detectable TDR reflection.

TDR signatures obtained with crushed gravel backfill are shown on the right side of Figure 7-10, where the initial three reference crimps were made in the cable prior to placing it in the CST. The two shear planes are represented by heavy horizontal lines and arrows in the CST schematic on the left side of the figure. Note the dual TDR reflections which developed at these shear planes as cable was bent above and below the upper and lower shear planes as shown in Figure 7-4b.

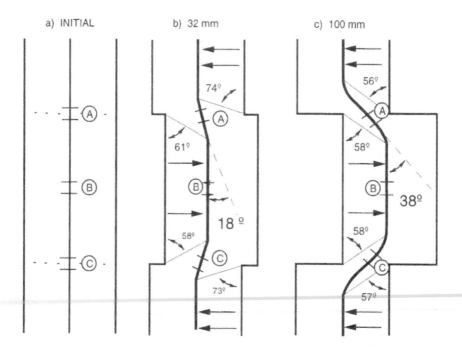

Figure 7-12. Approximate dimensions of passive wedges which developed in the gravel backfill during CST tests and representative cable deformation (from Logan, 1989).

Magnitude Versus Displacement

The reflection coefficient magnitude versus pipe displacement for tests conducted with dry gravel backfill are shown in Figure 7-13b. The stress-displacement calibration test results are shown in Figure 7-13a to illustrate that cable deformation was not detected until the peak shear strength of the gravel was exceeded and plastic deformation occurred.

The relationship between reflection magnitude and pipe displacement is essentially linear and this trend is consistent with the results presented in Figures 5-3 and 5-5 for tests with cement grout

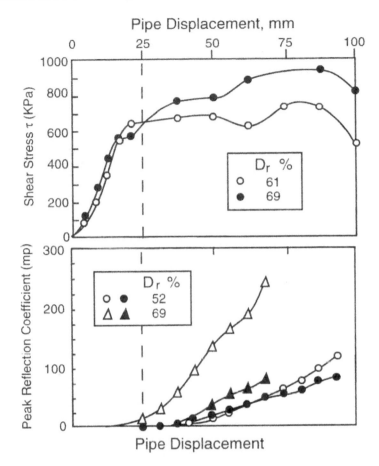

Figure 7-13. Compilation of several different CST tests:
(a) estimated shear stress; (b) TDR reflection magnitude. Note
that TDR reflections were not detected until plastic deformation
occurred in the gravel backfill; compare with Figure 5-3 (from
Logan, 1989).

backfill. The results are summarized in Table 7.11. The pipe
displacement shows no effect on the TDR signatures until displacement of 25 to 32 mm had occurred, which corresponds with the
limit of elastic deformation for the gravel.

  After all of the tests were conducted, the approximate
minimum radius of curvature was measured from the available
deformed cables or from photographs of cables. Cables deformed

**Table 7.11. Summary of test results with cable
in gravel backfill (Logan, 1989).**

Backfill	Number of anchors	Pipe displacement when cable failure occurred (mm)		TDR sensitivity (mρ/mm)		
		min	max	min	mean	max
dry gravel	1	32	38	1.6	4.4	6.3
dry gravel	3	25	32	1.5	3.3	6.0
wet gravel	1	32	44	0.4	1.6	3.1
sand	1			no reflectance changes		

just to the point of a detectable TDR reflection signature had a radius of curvature of 125 mm (5.0 in.). The dry gravel tests with one anchor had the cable fail with a minimum radius of 75 mm (3.0 in.). It can be inferred that no TDR reflection can be detected until the 12.7 mm Cablewave Systems coaxial cable is bent at a radius of curvature less than 125 mm (5.0 in.) as also noted by Su (1987).

Low Sensitivity of Cable with Gravel Backfill

Reflection magnitude for shear at one location along a short (less than 5 m long) cable is plotted versus displacement for different backfill types in Figure 7-14. The correlation between TDR reflection magnitude and shear displacement is assumed to be linear for these conditions, and there is a minimum displacement which must occur before any detectable reflection is created. It is most important to employ a grout backfill that is as compliant as the soil being monitored, so that it will deform locally if the soil does so.

Monitoring of cables embedded in granular material subjected to shear in a large diameter direct shear device has led to insights into interaction between the backfill and embedded cable. When the material was subjected to direct shear, a passive resistance plug formed, and displacement of the direct shear device was

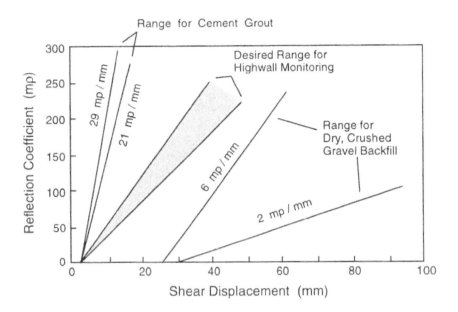

Figure 7-14. Summary of laboratory direct shear calibrations for 12.7 mm diameter Cablewave Systems smooth-wall aluminum coaxial cable in steel pipe.

transmitted to the embedded cable by this cone-shaped plug. The initial mode of cable deformation was bending with subsequent tensile deformation within the backfill shear zone. When the backfill friction angle was high enough, the cable was deformed over a relatively short length and the radius of curvature was small enough to produce a detectable TDR reflection. Furthermore, when the backfill was saturated, the interface friction between backfill and cable was significantly reduced so that even though the cable was deformed, it never failed in tension.

The testing revealed that it was feasible to use TDR for monitoring shear using 12 mm crushed gravel as the backfill material in a 200 mm borehole. It is unlikely that sand passing the No. 4 Sieve (4.75 mm) would ever be suitable. Anchoring the cable with cable clamps had no positive benefits when crushed gravel was used as the backfill.

Chapter 8
MONITORING STRUCTURAL DEFORMATION

This chapter presents the use of both metallic cable and optical fiber TDR to monitor structural deformation. While optical fiber TDR (OTDR) is not the focus of this book, a brief introduction to its physics and use is presented. Like metallic TDR (MTDR), OTDR-based instruments can be employed to monitor large volumes with a single cable/fiber. However, because of the differences between optical fiber and metallic cables, optical fiber allows measurement of response to small deformation, namely, elastic stress and strain. The chapter begins by describing use of MTDR for two aspects of structural response: 1) internal deformation, which involves monitoring localized response of individual structural members, and 2) external deformation, which involves monitoring the overall movement of a structure. The chapter then closes with an overview of OTDR measurement of stress and strain within a structure.

MTDR VERSUS OTDR FOR STRUCTURAL MONITORING

As discussed in Chapter 2, MTDR involves launching pulses of electromagnetic energy in the radio and microwave frequency range of 10^3 MHZ with wavelengths on the order of 1 m or less. For purposes of monitoring deformation of rock or structures, the transmission line consists of metallic coaxial cable which has mechanical properties very similar to engineering structural materials. In particular, metallic cable can deform plastically while still allowing transmission of electromagnetic energy, which makes it attractive for situations in which very large deformations are expected.

OTDR involves launching pulses of electromagnetic energy in the frequency range of 10^9 MHZ with wavelengths on the order of 10^{-6} m, although the light pulses themselves have lengths on the order of 1 cm. Optical fibers can be manufactured with Bragg gratings or reflective defects of various wavelengths (Wanser et al., 1994; Kersey, 1994) and microbend devices can be attached (Davis et al., 1986) to monitor elastic structural deformation. OTDR systems are not susceptible to electromagnetic noise, signals can be transmitted over distances of several kilometers, and optical fiber is much smaller and lighter than coaxial cable. Since optical fiber is stiffer and more brittle than metallic cable, it is suitable for monitoring deformation in the elastic range, and the wavelengths used with OTDR make it the only choice for this range of deformation.

Advantages and limitations of OTDR versus MTDR must be evaluated in light of the objectives of an instrumentation program. This chapter focuses upon examples in which MTDR has been used for monitoring large deformation and cracking of structures. Examples in which OTDR has been used for measurement of elastic deformation are presented in an extensive volume of literature, and two sources which offer a wide range of insight into uses of OTDR for structural monitoring are O'Connor et al. (1994) and Ansari (1993). At the close of this chapter, uses of OTDR to measure elastic stress and strain will be summarized, because MTDR has not been employed in such cases.

INTERNAL STRUCTURAL DEFORMATION— INSTALLATION AND PERFORMANCE

One of the major reasons that TDR is attractive for monitoring the deformation of rock and soil is that the cables can be easily embedded in, and fixed to, the mass by field personnel. In this chapter, two cases are presented to demonstrate installation procedures and performance of MTDR for monitoring structural deformation. These cases were attractive applications because of the ease of installation.

Masonry Wall on Spread Footing

In the fall of 1995, the U.S. Bureau of Mines constructed a prototype to evaluate a variety of techniques for constructing residential foundations which would be resistant to damage from subsidence-induced curvature (Triplett et al., 1995). The 6 m x 9 m (20 ft x 30 ft) foundation consisted of concrete block on a concrete spread footing. The footing was reinforced along the north and east walls, but not along the south or west walls, as summarized in Table 8.1. Similarly, reinforcement was placed within the concrete block along the north and east walls. A wooden deck was constructed on the foundation and residential loading was simulated by placing piles of sand on the deck.

Table 8.1. Cables installed in masonry block foundation.

Wall length	Footing type	Cable no.		Wall type	Cable no.
North - 6 m	reinforced	top	NFT	bond beam	NBB
		bottom	NFB		
East - 9 m		top	EFT		EBB
		bottom	EFB		
South - 6 m	not reinforced	top	SFT	n/a	
		bottom	SFB		
West - 9 m		top	WFT		
		bottom	WFB		

Embedded cable:
  Corrugated copper, 12.7 mm diameter, 50 ohm, Cablewave Systems FLC12-50.
Lead cable:
  Braided copper, RG11, low loss, CommScope F1160BVV.
  Braided copper, RG8, Type 50 ohm, Belden T8214 1C11.
  Braided copper, RG8/V, Type Coax HS, 50 ohm, Belden 9913.

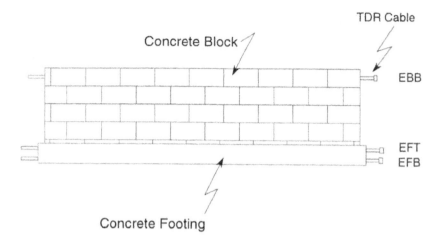

Figure 8-1. Coaxial cables installed in the footing and in the top course of concrete block, along the east wall of the foundation.

Installation

Vertical displacement and tilt (differential vertical displacement) of the foundation were monitored with an optical level, a tiltmeter, and taping from a string baseline. Strains were monitored externally with a tape extensometer and internally via TDR monitoring of 12.7 mm corrugated copper coaxial cables (Cablewave FLC 12-50) embedded in the footings and top course of blocks as shown in Figure 8-1. Visual crack surveys were conducted throughout the test.

Performance

Curvature of the foundation induced by ground movement was simulated by installing air bags beneath the footing at various points. The bags were inflated in a variety of sequences to simulate constant tilt and differential tilt (constant curvature) along the walls. As the curvature increased, large vertical cracks developed in the unreinforced walls and footings. However, only hairline cracks developed in the reinforced walls.

    During the test, which was conducted in Minneapolis in

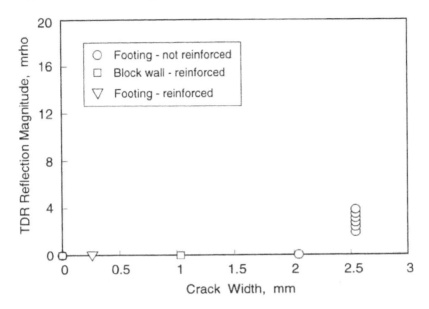

Figure 8-2. **Compilation of TDR reflection data for cables installed in the footings and walls of the prototype masonry foundation.**

November, problems developed with the TDR data acquisition system due to the low temperature and low battery power, and it was necessary to collect data using the strip chart recorder option available with the Tektronix 1502C cable tester. Due to this limitation, it was possible to acquire waveforms only during periods when a complete suite of manual measurements was made.

Detectable TDR reflections developed at locations where crack widths were greater than 2 mm and reflection magnitudes are compiled in Figure 8-2. These larger cracks appeared in the unreinforced footings along the south and west walls. This sensitivity is slightly better than the findings of Panek and Tesch (1981). They used 12.7 mm diameter smooth-walled aluminum cable (FXA12-50, Cablewave Systems) and found that detectable reflections developed only for crack widths of 6 mm (1/4 inch) or greater. Obviously, the performance is enhanced with corrugated copper coaxial cable. It is important to note that the cracking developed in response to curvature-induced tensile strain. A more consistent approach would have been to adopt the technique employed by Aimone-Martin et al. (1994) where the change in distance between reference crimps in the cable was measured as summarized in Table 5.2.

Reinforced Concrete Column

TDR cables may offer a means of quickly detecting material cracking damage of reinforced concrete structures caused by earthquake shaking. The cables could be left in place for decades and then probed after earthquakes. Among the more critical structures are bridge piers that support elevated sections of roadway. In a recent laboratory study, lateral loading of prototype bridge piers revealed that tensile deformation occurs at the interface of the column and its base, and that vertical slippage or shearing occurs between overlapped, or spliced, steel reinforcement within the column (Lin, Gamble, and Hawkins, 1994). Splicing is a common construction practice that provides a weakness when a splice is located near large moments that develop within a column during an earthquake. For this reason, most bridge columns in seismic zones are being retrofitted with external reinforcement along the spliced lengths to provide additional moment resistance.

Installation

To study the usefulness of TDR cables, prototype reinforced concrete bridge columns were fitted with vertical and transverse TDR cables at the base near reinforcement splices and subjected to cyclic lateral loading. As shown in Figure 8-3, columns were 3.66 m (12 ft) high and 0.6 m (2 ft) in diameter, and vertical steel reinforcing bars were spliced through the bottom 1 m (3 ft) of each column. Two cables were vertically aligned to primarily monitor tensile deformations at the column-base interface, and two cables were installed transversely through the column diameters perpendicular to the vertical cables to monitor shearing between the spliced reinforcement. The dashed lines in Figure 8-3 mark cable segments encased in concrete; solid lines represent the exposed lengths of cable. Each vertical cable passed through the base beam and exited on one side. Both ends of each transverse cable were exposed to allow interrogation from either end with a TDR pulser.

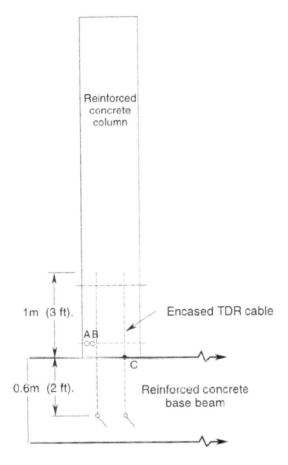

Figure 8-3. Coaxial cables installed in a prototype reinforced concrete column (from Pierce and Dowding, 1995).

Dowel reinforcement consisted of #11 (1.38 in. or 3.50 cm diameter) steel bars that extended up from the base beam and into the bottom 0.91 m (3 ft) of the column. Column reinforcement also utilized #11 steel, but these bars were spliced with steel wire ties to the dowel bars where they overlap in the bottom fourth of the column. Column splicing is a common field construction practice, and this zone is often the weakest section of a laterally loaded column.

The cable employed for TDR monitoring in shear-sensitive regions consisted of a smooth-wall aluminum shield with a solid copper-clad aluminum inner conductor, separated by a polyethylene foam dielectric (Cablewave FXA12-50). The cable employed in extension-sensitive regions had the same inner conductor and dielectric material, but was surrounded by a corrugated copper shield (Cablewave FLC 12-50). Both cables are commercially available without a protective jacket. These cables are robust yet flexible for easy installation. For simplicity, the two cables are referred to as "shear" and "extension" cables.

One end of each vertically oriented extension cable was encased in the column, as shown in Figure 8-3. As a precaution, the metallic conductors at this end were sealed to prevent infiltration of water and to mitigate cable degradation. A plastic tip, 12.7 mm diameter and 19 mm long, was placed over the end and taped to the cable.

All cables were spray painted before installation to deter chemical interaction. It has been shown that aluminum corrodes in a high pH (approximately 12) environment of curing concrete before the water becomes immobilized (Kim, 1989). Even for the aluminum, the interaction declines greatly when water becomes fixed after curing. Therefore, a thin coat of paint suffices to insulate the cables until the concrete has set. There was uncertainty as to the susceptibility of copper to corrosion in curing concrete, so the copper cables were painted as a precautionary measure.

Two extension TDR cables were attached to the dowel reinforcement by thin, steel wires. For long-term field conditions, plastic ties should be employed to avoid establishing an iron–aluminum or iron–copper corrosion cell. Each cable was easily bent by hand through a 90 degree arc with a 0.40 m (1.3 ft) radius of curvature inside the beam to allow a perpendicular exit from the beam and form work. A continuous, non-kinked curve does not produce a voltage reflection and so will not interfere with TDR monitoring (Pierce et al., 1994).

Performance

The columns were individually tested under cyclic lateral loading to simulate earthquake conditions. Essentially, the free end (top) of the column was displaced laterally in controlled cycles, creating large bending moments at the column-base interface. Displacements were increased incrementally over time to failure, defined as a post-peak loading equal to 80% of the maximum measured load. A more thorough discussion of the testing procedure is given by Lin, Gamble, and Hawkins (1994).

Cable deformation during cyclic loading was measured by interrogating cables when the column was at the maximum positive and negative displacement for each cycle. TDR waveforms were collected and analyzed immediately. Not all of the cables were deformed sufficiently to produce detectable reflections, but several cables provided excellent response. In general, fracturing of the concrete subjected the vertical cables to extension at the column-base interface, and subjected the transverse cables to shear where reinforcement bars were spliced together and debonding from the concrete occured.

In one column test, shear deformation was detected on the lower transverse cable at points A and B, shown in Figure 8-3. Tensile deformation occurred along one of the vertical cables at the column-base interface in another test, designated as point C. Reflection coefficients measured from these cable deformations are compared to simultaneous strain gage measurements of the nearest reinforcing bars in Figure 8-4. The greater slope for shearing of the transverse cables indicates the greater sensitivity of TDR to this mode of cable deformation. Tensile strain measured in the reinforcing steel does not necessarily correlate with cable shearing or extension, so a direct comparison cannot be made.

Interpretation of the TDR response is more meaningful when compared with the load-deformation behavior of the column. Figure 8-5 shows the *maximum* load and displacement points for column D for each *cycle* of loading, in the positive and negative directions. Displacement measurements were taken at the top of the column, meaning these displacements are the maximum sustained by the column. The hatched band represents the area between global maximum load in the two directions.

Also shown in Figure 8-5 are the responses of two transverse shear cables, one located 1 m (3 ft) above the base (upper) and one located 0.15 m (0.5 ft) above the base (lower). It should be expected that damage is concentrated in the region where

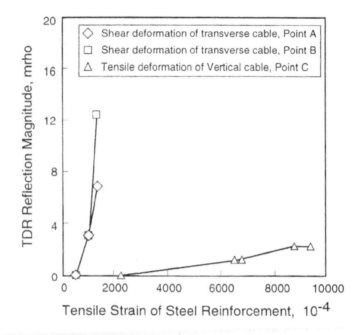

Figure 8-4. Compilation of TDR reflection data for cyclic lateral loading of prototype column; point locations are shown in Figure 8-3 (from Pierce and Dowding, 1995).

the column is connected to the base. The greater deformation of the lower cable is consistent with this behavior, as indicated by reflection coefficient magnitudes 4 to 12 times greater than those measured on the upper cable. Also, deformation of the lower cable produced detectable TDR reflections one cycle before deformation was detected on the upper cable.

Detectable TDR reflections did not develop until after global peak loading was achieved. This suggests that the column was behaving elastically up to peak load, such that no permanent (plastic) damage or cracking occurred. When the cables were deformed, this occurred at locations within the reinforcement lap splice. This suggests that cable deformation was initiated as the reinforcement strained sufficiently to break the bond between steel and concrete similar to the results obtained by Huston et al. (1994), who used optical fibers to monitor debonding.

Response of vertically oriented cables is shown in relation to the load-deformation response of column F in Figure 8-6.

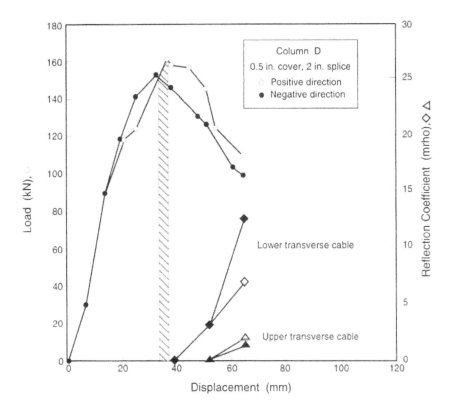

Figure 8-5. Development of shear deformation of transverse cables as debonding occurred between the reinforcement and concrete during cyclic lateral loading of Column D. Data points correspond with the *maximum* load and displacement for each *cycle* of loading.

Column F was similar to column D except there was more concrete cover over the exterior reinforcing bars. A detectable TDR reflection developed in one of the vertical cables during post-peak loading. The location of this reflection corresponds to a point on the cable at the column-base interface. The TDR reflection increased in magnitude only when the cable was extended (i.e., when the column was displaced in the positive direction). Further, a more consistent approach for cable extension would have been to adopt the technique employed by Aimone-Martin et al. (1994) where the change in distance between reference crimps in the cable

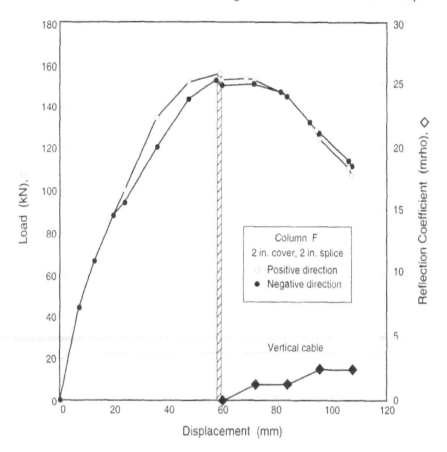

Figure 8-6. Development of cable extension at column base during cyclic lateral loading of column F. Column F was similar to Column D except for increased concrete cover over reinforcement.

was measured as summarized in Table 5.2.

EXTERNAL STRUCTURAL DEFORMATION— INSTALLATION

Scour has been linked to nearly 95% of all severely damaged and failed highway bridges constructed over waterways in the United States (Lefter, 1993). There are two issues associated with such scour-induced damage to bridge pier footings. The first effect is the loss of foundation material, which exposes the footing and lowers its factor of safety with regard to sliding or lateral deformation.

The greatest loss of sediment to scour occurs at high water velocities, such as during floods. Second, pier movement may occur as a result of material loss beside and beneath the base of the footing, which produces undesired stresses in the bridge structure and ultimately results in structural collapse. For these reasons, the study of bridge scour has drawn much attention in recent years.

TDR technology can be deployed to detect bridge scour and monitor pier and abutment movement during flood events, when the maximum amount of scour is likely to occur. However, application of TDR to bridge foundation monitoring requires development of an innovative means to package and install the cable system underwater. As shown in Figure 8-7, the envisioned TDR scour detection system measures the length of a buried cable, where flanges attached to the cable are expected to be torn off by high drag forces as the river bed is scoured away, thus shearing the cable at the flange connection. The shearing process shortens the cable and a signal reflection produced at the cable break marks the scour depth. As shown in Figure 8-8, measurement of lateral pier or abutment movement is accomplished with a similar system without flanges. As the footing displaces relative to the underlying rock or soil, the cable is deformed and a TDR reflection is created at the location of distortion. The reflection magnitude is directly proportional to the amount of cable deformation, which, in turn, corresponds to the footing displacement.

Scour Depth Measurement During Flood

The proposed design of a TDR system for measuring scour depth is based on recording voltage reflections arising from shear deformation at known weakness locations along a vertical, metallic coaxial cable embedded in front of a bridge pier as shown in Figure 8-7. Flanges are positioned at these weakened locations to precipitate shearing of the cable. When the river bed erodes to a flange elevation, it will be exposed to the high water velocity environment. The water drag forces will load the flange and deform the cable. As the cable deforms, a TDR reflection will develop and indicate scour has occurred at the flange elevation.

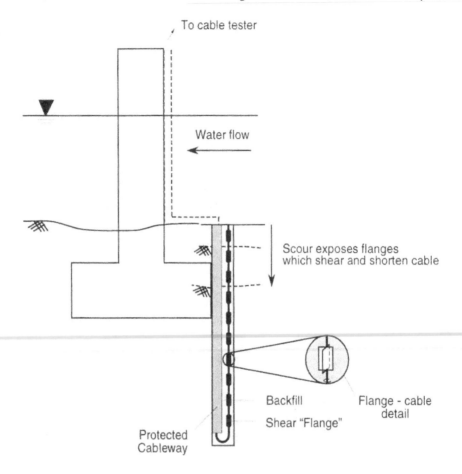

Figure 8-7. Preliminary design of cable installation to monitor scour beneath a pier foundation during floods (from Dowding and Pierce, 1994b).

The most challenging aspect of this design is the installation of the cable and flanges in front of a pier footing. Placement of the device is envisioned on the upstream side of a pier and in front of the spread footing. A borehole would be drilled at this location and cased to allow installation of the preconstructed cable system. The cable system will consist of a single metallic cable placed vertically inside a protective conduit to its open end, where the cable is bent 180° to extend upward alongside the conduit. Vane-shaped shear flanges would be attached to the cable at predetermined locations. The single cable with multiple flanges would then be lowered into the cased hole, after which the hole should be carefully backfilled with a clean sand. As the backfill is tremied to the existing river bottom elevation, the casing is removed, leaving the exposed length of cable with the attached flanges buried in a sand column adjacent to the footing. As scour occurs near the cable, the backfill will be removed along with the natural bed materials. Exposing the cable will subject a flange to high velocity water flow, thus causing flange rotation and subsequent shearing of the cable. This operation is depicted in the detailed drawing of the flange shown in Figure 8-7.

An alternative scheme for measuring bridge scour using TDR involves locating the river bed elevation by measuring voltage reflections due to changes in cable impedance at the interface of the saturated soil and flowing water. This method invokes a technique from research on water level detection which is discussed in Chapter 9.

Bridge Pier or Abutment Movement

The compliant cable-grout system described in Chapter 7 can also be applied to monitor footing or abutment movements resulting from scour and other processes. Figure 8-8 illustrates a cable extending from the cable tester, down along the pier, and into a hole drilled through the footing and foundation material. The cable is encased in grout from the bottom of the hole to the top of the footing. A protective pipe, needed to screen the cable from debris, encloses the cable from the top of the footing to the bridge deck. Lateral footing movement would produce localized shearing of the grouted cable at the interface of the footing and foundation soil. Cable deformation would occur after the grout fractured and this deformation would produce a detectable TDR reflection as discussed in Chapters 5 and 7.

Lateral translation or rotation of the footing would progressively deform the cable, thereby producing a TDR reflection. Shearing deformation of metallic coaxial cable would allow detection of pier movements on the order of millimeters, which would be sufficiently sensitive to initiate an alarm.

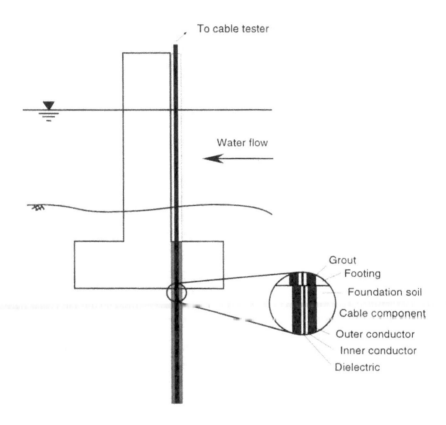

Figure 8-8. Preliminary design of cable installation to monitor lateral movement of bridge pier (from Dowding and Pierce, 1994b).

OTDR AND ELASTIC STRUCTURAL DEFORMATION

Use of optical fiber for TDR measurement of structural deformation greatly broadens the application spectrum into measurement of elastic behavior. As with MTDR, OTDR combines transmission and measurement in the same fiber, and has the potential to monitor large volumes. Properties of optical fiber allow measurements to be made which are different from those that can be made using metallic cable. Two such measurements are longitudinal elastic strain and cyclic stress. This section will focus on these two measurements after a brief background on cable and measurement systems. This treatment is very cursory, but should be sufficient to alert the reader as to the differences between MTDR and OTDR measurement approaches.

Fiber Optic Component of a System[1]

Fiber optic cables are waveguides that carry electromagnetic energy in the visible and infrared portion of the spectrum. Most cables are glass, with plastic cables used in some limited applications. Looking at the end of the cable, highly magnified, one can see the core, cladding, and buffer as shown in Figure 8-9 (Sterling, 1987). The core carries the light signal. Core size varies in different cable types as noted in Table 8.2. Single-mode cable, used in long range and high bandwidth systems, has a core size of 8 to 10 µm. Multi-mode cable, used in shorter range, medium bandwidth applications, has a core size of 62.5 µm. Plastic fibers, for very short range, low bandwidth applications, has a core size of 200 µm. The cladding, with an outside diameter of 125 µm, is glass similar to the core but with a slightly different composition. This layer keeps light in the

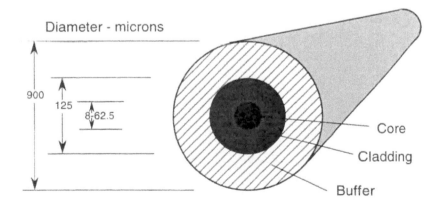

Figure 8-9. Fiber optic cable construction (from Smith, 1994).

---

[1] This section is reproduced from Smith (1994).

Table 8.2. Cable and optical wavelength combinations (from Smith, 1994).

Wavelength (nm)	Source type	Core diameter[1] (μm)	Maximum range (km)	Bandwidth/ resolution[2]	System characteristics
650-780	LED	50 - 200	0.1	very low	very inexpensive
850		50 - 62.5	1	low	inexpensive
1300	LED/Laser		3	medium	inexpensive, bandwidth
1310	Laser	8 - 10	20		long range, bandwidth
1550			150	high	range, sensitivity

[1] Plastic fiber core diameter is 200 μm; glass fiber core diameter is 100 μm for large core, 62.5 μm for commercial multimode, 50 μm for multimode, and 9-10 μm for single-mode cable.

[2] Bandwidth/resolution varies by the distance the signal must travel. The further the signal travels, the lower the bandwidth, often specified as Mbit or MHz per km.

[3] Higher wavelengths, particularly 1550 nm, are more sensitive to stress and bending.

core and prevents external light from entering the core. Surrounding the cladding is a buffer layer that protects the inner glass. The buffer outside diameter is 900 µm and it is made of ceramic or plastic.

Light is launched into the cable from a LED (Light Emitting Diode) or laser transmitter. LEDs are lower wavelengths of light, 600 to 850 nm in the visible spectrum and near infrared. They are inexpensive and used in low bandwidth applications. Lasers transmit 1300 to 1550 nm wavelengths in the far infrared, and are more expensive. They are used in longer range and higher bandwidth applications. The transmitter acts as an electrical to optical converter. Different wavelengths are used on the different cable types. Common combinations of wavelength and cable are shown in Table 8.2.

OTDR Component of a System

The Optical Time Domain Reflectometer (OTDR) consists of a transmitter laser, a receiver, and a time measurement computer controlling the two devices. The laser launches a pulse of light into the cable. The receiver samples the cable at periodic intervals, looking for reflections. The time between the initial pulse and the reflection relates to the distance to an event on the cable. The amount of reflection from the event gives information on the type of event and its level of effect on the cable. There are three different types of OTDRs: Fresnel, backscatter, and photon-counting (Tektronix, 1992). The differences are a) the types of reflection that can be seen, b) the distance resolution, and c) the distance range.

An interface between mediums with different refractive indices will reflect or refract light depending on the angle of the light beam (Sterling, 1987). The glass in fiber optic cables has an index of refraction, n_{glass}, of approximately 1.5. When light traveling through the glass fiber hits the end of the glass and encounters air ($n_{air} = 1.0$), the interface creates a large reflection called a Fresnel reflection. A second type of reflection comes from the glass itself. Small impurities and the crystalline structure of the glass cause a small amount of the light beam to scatter. This is Rayleigh scattering (Sterling, 1987). Some of the scattered light returns down the fiber to the OTDR and the returned light is called backscatter (Tektronix, 1993).

Light is received by an optical-to-electrical diode receiver. Two different types of diodes are used depending on the application. PIN (Positive-Intrinsic-Negative) photodiode receivers are used for high intensity, high bandwidth light signals. APD (Avalanche Photo Diode) receivers are used to detect lower bandwidth, very low power signals.

Fresnel reflections allow high resolution measurement of the distance to major refractive index interfaces. Fresnel OTDRs use a PIN diode receiver and very narrow light pulses, which permits distance resolution to better than 1 mm (Tektronix, 1992). While the narrow pulses allow high-resolution distance measurements, it also limits the range to about 100 m. Fresnel OTDRs are limited to sensors and cable events that create Fresnel reflections. The Tektronix CSA803 with optical plug-ins is an example of a Fresnel OTDR.

Backscatter OTDRs use more sensitive APD receivers that can detect very small amounts of reflected light. These reflections allow the OTDR to see changes that affect the cable itself, such as stress, bending, kinks, temperature, etc. The tradeoff is that very large reflections from Fresnel events can saturate the receiver, reducing resolution. Also, large pulses of light are launched into a cable to provide a strong backscatter signal which further reduces distance resolution. A backscatter OTDR using wide pulses can be used to interrogate very long sections of cable, up to 150 km, but the distance resolution is only 2 to 10 m (Tektronix, 1993). The Tektronix TFP2 is a backscatter OTDR.

Photon-counting OTDRs are a variation of backscatter OTDRs. These OTDRs use a receiver optimized to see smaller reflections than the backscatter OTDR. The more sensitive receiver allows the use of narrow test pulses for distance resolution, similar to Fresnel OTDRs with the ability to detect backscatter. The drawbacks are cost, two or more times greater than other OTDRs, and very limited range. An example of a photon-counting OTDR is the Opto-Electronics PPC series (Opto-Electronics, 1994).

Other Optical Measurement Techniques

OTDR is an event-specific method of testing optical components. One can see individual cable events separately on the display. Other testing methods measure changes in the entire cable by passing a signal through the cable, or by looking at the entire reflection at once.

Power-loss testing can detect if bends and kinks have occurred on the fiber, but cannot locate an event or detect if there is more than one event. It can be used as a low cost method to see if changes have occurred, followed by an OTDR test to find the exact location of a change. A light transmitter is placed at one end of the fiber, a detector at the other, and events along the fiber will change the level of light received (He and Cuomo, 1991). Power-level testing will detect effects similar to a backscatter OTDR.

Return-loss meters measure the amount of light reflected from a fiber. Like power measurements, they inject light and measure the output, but they measure reflected light at the same end as the transmitter. They are used to measure changes in overall backscatter and end reflection (He and Cuomo, 1991).

A method similar to power-level testing uses polarized light to detect very small changes in the fiber. Polarized light is transmitted into the fiber, and very small changes can be measured by checking the polarity of the light received. This can be used for high resolution stress and radiation measurements. Polarization measurement can detect very small, short-range effects similar to photon counting OTDRs (Dandridge et al., 1984).

MEASUREMENT OF STRESS[2]

External stimuli such as applied stress can influence light traveling through fiber optic cable and thus light transmission can be correlated with stress level. External compressive stresses can be used to introduce small strains in the fiber, which, in turn, induce a

---

[2] This section is reproduced from Dubaniewicz et al. (1994).

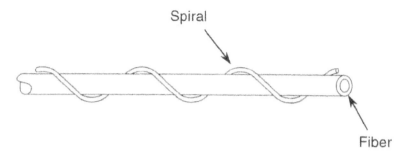

Figure 8-10. Unprotected Hergalite spiral-wound optical fiber (from Dubaniewicz et al., 1994).

loss of light. These light losses, caused by small amplitude, high-spatial-frequency perturbations, are called microbending losses (Allard, 1990). Microbending losses also affect the Rayleigh backscatter signature of fiber optic cable (Measures et al., 1991; Mickelson et al., 1984). OTDRs can continuously detect changes in the Rayleigh backscatter along a length of fiber optic cable. By defining sensing regions spaced several meters apart, interference between successive sensing locations along the fiber will be low (Mickelson et al., 1984). OTDR has also been used to measure stress with a Brillouin backscatter technique (Horiguchi et al., 1992) continuously.

Candidate Fibers

The following example involved the use of transmitted, rather than reflected, light. However, the approach used by Dubaniewicz et al. (1994) to develop a stress sensor illustrates how optical fiber can be utilized for this application. First, they identified several types of stress-sensitive optic fiber cables as potential candidates for a stress sensor. One commercial fiber has a microbending mechanism built into the fiber's construction. This cable, know as Hergalite, shown in Figure 8-10, consists of an optical fiber surrounded by a plastic spiral-wound strand. Plastic tape covering the Hergalite provides a surface for pressing the spiral-wound strand into the fiber. As the cable assembly is compressed, the plastic strand imparts microbends into the optical fiber, reducing the fiber's ability to transmit light.

    Another candidate fiber is manufactured by SpecTran

Corporation. This fiber is specially doped to make it more sensitive to applied pressure. The sensitivity is such that the fiber does not require any special microbending mechanism. Simply squeezing the fiber is enough to attenuate the light signal. The basic fiber is available with a number of different coatings, including a normal coat, an "F" coat, a silicone-based coat, and a Teflon-based coat. The attenuation of this fiber is significantly higher than communications-grade multimode fiber. For instance, the attenuation of the silicone-coated fiber is reported to be approximately 15 dB/km at 850 nm and 23 dB/km at 1300 nm, whereas communications-grade fiber is on the order of a few tenths to a few dB/km at these wavelengths.

Direct Transverse Strain and Embedded Stress Tests

To gain a basic understanding of the sensitivity and linearity of response of the various stress-sensitive cables to direct transverse strain, optic fibers were compressed in a simple displacement jig. This jig consisted of a micrometer head accurate to 0.00254 mm (0.0001 in.) used to control 2.54-cm (1 in.) thick steel platens. To perform a direct transverse strain test using this jig, one end of a cable was connected to a light source and the opposite end to an optical power meter. The cable was then inserted between the jig's platens, effectively creating an active gage length of 2.54 cm. Then a micrometer was used to minutely compress the cable between the steel platens in increments of 0.0254 mm (0.001 in.). Cable strain was calculated as the change in cable diameter divided by the original diameter.

The cables showed a wide range of sensitivity to transverse strain. The spiral-wound (Hergalite) fiber showed the greatest sensitivity, with almost all of the light eliminated at approximately 20% strain. The sensitivity of the various SpecTran cables appeared to be directly related to the relative thickness (stiffness) of the specific coating. The decrease in the light-loss sensitivity of the SpecTran cable with increasing thickness/stiffness of the coating suggests that the thicker coatings absorb more of the cable strain and actually shield the light transmitting fiber core from microbending losses. The range of light-loss sensitivities offered by

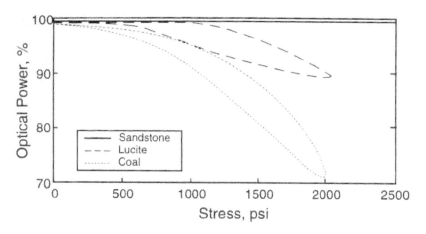

Figure 8-11. Results for the embedded stress tests with spiral-wound optical fiber grouted into various materials (from Dubaniewicz et al., 1994).

these various cables makes it possible to custom design stress sensors for a wide range of materials.

The relative stiffness of grout employed to fuse the optic fiber and surrounding structural or geologic material is important. To measure this effect, cables were grouted into blocks of materials with a wide range of stiffness and then the blocks were subjected to gradual loading and unloading. A 1300 nm LED was connected to one end of the fiber and an optical power meter was connected to the other end.

Several combinations of materials were used in these embedded stress tests: Sulfaset, Lucite, coal, and sandstone. Sulfaset is a quick-setting sulfur-cement-based material typically used to grout machinery mounts and roof bolts. In these tests, Sulfaset was used to grout the fiber optic cables into a borehole in the various materials, since the elastic modulus of Sulfaset is similar to that of coal. Hypothetically, a stress-sensitive fiber optic cable grouted into a coal seam using Sulfaset would be subjected to approximately the same magnitude of stress as the surrounding coal. Lucite was chosen for a similar reason. The modulus of Lucite [about 2,827 MPa (410,000 psi)] is close to that of coal in the Pittsburgh Coal Seam. Coal was also used, although not extensively because of the crumbling problem, to get the most realistic results. Sandstone was also tested to see how the fiber would respond in a much stiffer material.

Figure 8-11 shows the transmitted power response of Hergalite spiral-wound fiber embedded in these materials. The sandstone sample showed no response until it failed. It is much stiffer than the Sulfaset and the load simply arched around the embedded cable. Experiments with the Lucite/Sulfaset and coal/Sulfaset samples showed only a slight change in the light transmitted at lower stress levels, then a moderate change as the compressive stress increased above 7 MPa (1000 psi). There was some hysteresis evident in both cases as well. The hysteresis tended to decrease as the number of trials increased. The Hergalite survived repeated testing of these two samples, and the response was generally repeatable. The SpecTran showed similar results.

MEASUREMENT OF LONGITUDINAL STRAIN[3]

Fiber Bragg gratings (FBGs) allow measurement of strain at multiple predetermined locations along an optic fiber. Bragg gratings can be written into Germanium doped optical fiber by exposing the fiber to an ultraviolet (UV) interference signal generated either holographically via two-beam interferometry (Meltz et al., 1989) or by using a diffraction mask (Hill et al., 1993). The absorption of UV light in the fiber changes the chemical bonds in the glass (producing defect centers), thus giving rise to a change in the complex refractive index of the glass. The gratings and the measurement concept are shown in Figure 8-12. The grating reflects light strongly at the wavelength, λ_B, for which the Bragg resonance is satisfied. This Bragg resonance is determined by the condition

$$\lambda_B = 2n\Lambda \qquad (8\text{-}1)$$

---

[3] This section is reproduced from Kersey (1994).

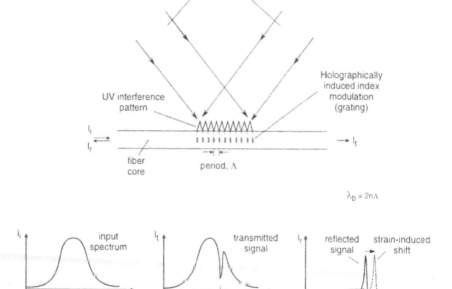

Figure 8-12. Fiber Bragg grating transmission and reflection spectra (from Kersey, 1994).

where Λ is the spatial pitch of the grating, and n is the effective index of refraction for the fiber. Strain applied to the grating thus shifts the wavelength at which the resonance condition is satisfied. Illumination of the grating using a broadband optical source thus produces a narrow-band reflection signal off the grating, the wavelength of which is encoded by the deformation being measured (Morey et al., 1991).

This inherent wavelength-encoded nature of the output of FBGs has a number of distinct advantages over other sensing schemes. One of the most important is that sensed information is encoded directly into the wavelength (which is an absolute parameter). The output does not depend directly on the total light levels, losses in the connecting fibers and couplers, or source power. Furthermore, the wavelength-encoded nature of the output also facilitates wavelength division/multiplexing by assigning each sensor to a different slice of the available source spectrum.

The key to a practical sensor system based on FBGs lies in the development of instrumentation capable of determining the relatively small shifts in Bragg wavelength of FBG elements induced by strain or temperature changes in these sensor elements, and in the low cost fabrication of the fiber grating elements (Askins et al., 1994). This former area has received significant attention, with a variety of approaches demonstrated. The range of application areas for FBG sensors could be quite extensive, but most interest is being directed at the development of quasi-distributed, multipoint, strain measurement systems for use as embedded sensors in structural sensing applications.

Chapter 9
AIR–LIQUID INTERFACES

While field measurement of water content for irrigation currently drives the TDR market, there are other water-related measurements of importance: 1) pore water pressure near critical structures, 2) elevation of the ground water table, and 3) elevation of other water surfaces. These measurements are easily made using TDR because of the very large voltage reflection that occurs at the air-water interface when water rises in the annular space of air-dielectric coaxial cable. This large amplitude reflection allows a wide range of TDR pulsers to be employed in this measurement. Background for TDR measurement of fluid levels and discrimination of fluid types is developed in this chapter, and TDR methods are compared with other approaches. Example applications are presented including the use of TDR technology to detect leakage of a variety of liquids over large areas, or within large volumes. This measurement is possible with specially designed coaxial cables that allow liquids to penetrate the outer conductor and permeate the dielectric material.

BACKGROUND

Patents for use of TDR to measure liquid levels were first issued in 1974 (Ross, 1974; Ross, 1976), several years after such an application was proposed by Fellner-Feldegg (1969). As originally conceived, liquid levels in industrial containers could be measured by noting the TDR voltage reflection at the air–liquid interface inside hollow tubes inserted in the container. Because of clogging, the system eventually evolved to a single dielectric-coated line (i.e., Goubau line) coupled with a conductive container.

Dowding and Huang (1994b) adapted this same technology for use in field measurement of water levels of all varieties. While field measurement of water content by TDR is widely employed, field measurement of water levels by TDR offers an attractive alternative to other systems when it can be combined with other TDR applications (e.g., deformation monitoring), or when it offers advantages over these other methods. Water level measurements which are commonly measured in the field include:

- water levels in open channels (streams, irrigation canals, etc.),
- phreatic surfaces (e.g., water table, confined water levels, etc.) of the ground water, and
- pore water pressure in and around critical structures.

There are many situations in which these measurements must be made remotely or on a continuous basis, such as measurement of pore water pressure beneath the foundation and abutments of a dam to monitor stability.

TDR Reflection at the Air–Liquid Interface

TDR waveforms of reflection coefficient versus distance for a 50 ohm air-dielectric coaxial cable connected to a 50 ohm lead cable are shown in Figure 9-1a. The signal remains constant (point b to point d) until the end of cable is reached and a positive reflection (point d) occurs due to the open circuit. However, if the air-dielectric cable is partially immersed in water, a negative (downward) reflection (point c) occurs at the air–water interface because of a change in impedance (compare Figures 2-3a and 2-13).

The signal remains constant until the end of cable is reached (point f), where a positive reflection occurs. The cable appears longer when submerged in the liquid (point f versus point d), because the pulse propagation velocity, V_p, is reduced (distance = V_p x time $_{TDR}$) due to the greater permittivity of water (ϵ_{water} = K $_{water}$ ϵ_o). As discussed in Chapter 2, the dielectric constant, K, of air and water are approximately 1 and 81. Similarly, velocity would be different in benzene and ethanol, which have dielectric constants of approximately 2.3 and 24.3 (Kaya et al., 1994). V_p is the ratio of propagation velocity of an electromagnetic wave along a transmission line to the velocity of light in a vacuum (3×10^8 m/s). When

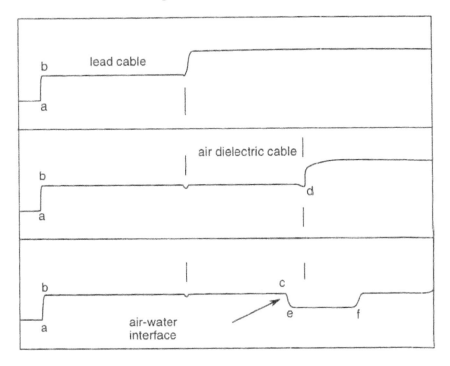

Figure 9-1. TDR waveforms showing reflection at air-water interface in air-dielectric coaxial cable: (a) waveform for 50 ohm lead cable; (b) waveform with 3 ft (1 m) long air-dielectric cable connected to lead cable; (c) air-dielectric cable submerged in 0.49 ft (0.15 m) of water; apparent submerged length, cf = 1.62 ft (0.49 m).

this cable is submerged in water, the apparent distance cf (= 1.62 ft) is greater than the actual submerged length, l_p = 0.49 ft. The apparent distance is also controlled by the propagation velocity setting of the TDR cable tester $V_{Ptester}$ (= 66%). For example, in Figure 9.1, $[(V_p)_{air} \times t] / [(V_p)_{water} \times t] = (2 \times cf) / [(l_p)(V_{Ptester})] = [(2)(1.62ft)]/[(0.49ft)(0.66)] = 10.0$ compared with $(K_{water})^{1/2}/(K_{air})^{1/2} = (81 / 1)^{1/2} = 9.0$ [note that cf is multiplied by 2 to account for the round trip distance of the TDR voltage pulse].

Sensitivity of a TDR waveform to changes in liquid levels within a tube is best summarized with work by Dowding and Huang (1994b). They conducted an investigation using twisted pair wire to determine the smallest diameter tubes that could be employed. Wires were inserted into a 3.2 mm (1/4-in.) inner diameter polyeth-

ylene, U-shaped tube partially filled with water. Response of
different wire types was investigated by visually comparing
characteristics of the TDR waveforms. Based on this investigation,
twisted pair wire with both conductors insulated was selected for
use in field simulation tests.

Field conditions were simulated to evaluate the effect which
variable rates of water level movement and residual water drops
have on measurement accuracy. Rapid water level rise, and both
rapid and slow water level decline, were simulated in an 8 m (25 ft)-
high standpipe. A water supply and valve was attached to the
bottom of the standpipe so water could be added or drained. The
twisted pair wire extended from the top to the bottom of the
standpipe.

Experiments started by incrementally raising the water level
in the standpipe and a baseline TDR waveform was obtained at the
maximum water height. Water was then allowed to fall at a rate of
15 m/day (0.15 MPa/day) to simulate unusually rapid dissipation of
water pressure, and a rate of 2 m/day (0.02 MPa/day) to simulate
less rapid dissipation of water pressure. These rates are much
greater than typical field conditions (approximately 0.001 MPa/
day). Slower rates of interface movement will allow evaporation of
water drops, which were found to have an influence on propagation
velocity of a twisted pair wire in a small diameter standpipe. Thus,
the more rapid rate of drawdown was the severest test of performance.

As shown in Figure 9-2, X_M is the physically measured
distance to the air-water interface, X_1 is the uncorrected TDR
transmission distance, and X_2 is a corrected transmission distance.
The shortest distance, corresponding to the most recent maximum
height of water, is X_D. The measured relationships between X_1 and
X_M for rising and falling water levels can be approximated by a 1:1
line within a few percent error, as shown in Figure 9-3. These X_1s
are determined on the basis of V_p in air only. It was not considered
necessary to account for the change in V_p produced by the presence
of water droplets in the tube; however, this procedure is discussed
below.

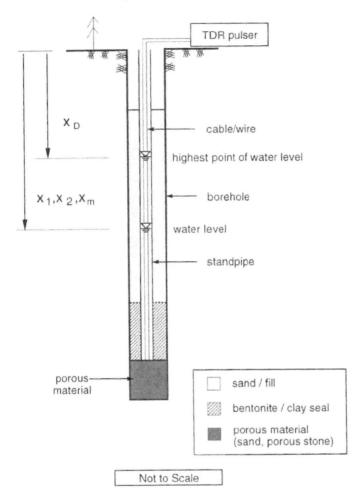

Figure 9-2. Schematic of installation for monitoring water level in a standpipe piezometer.

Correction for the Presence of Residual Water Drops

If it is found necessary to correct for changes in V_p due to the presence of residual water drops, the corrected transmission distance, X_2, can be found via Equation 9-1,

$$X_2 = X_D + \left(\frac{X_1 - X_D}{V_{air}} \right) V_{bubble} \qquad (9\text{-}1)$$

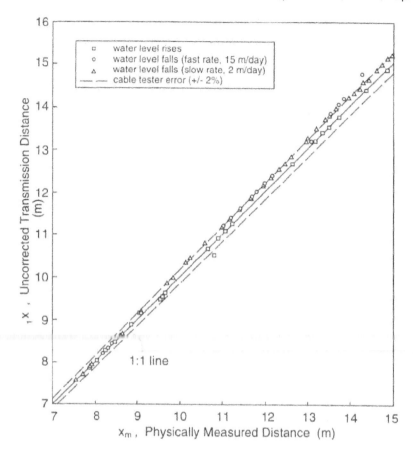

Figure 9-3. Correlation between TDR and physical measurement of distance to air-water interface (from Dowding, Huang, and McComb, 1996).

where V_{air} is the propagation velocity of the wire or cable in air, and V_{bubble} is the propagation velocity when it is surrounded by residual water drops. Rearranging Equation 9-1 and setting $X_2 = X_M$, V_{bubble} can be obtained by

$$V_{bubble} = \left(\frac{X_M - X_D}{X_1 - X_D} \right) V_{air}. \qquad (9\text{-}2)$$

For the twisted pair wire used in this study, V_{bubble} converged to a value of 0.72 with increasing transmission distance (compared with $V_p = 0.75$ in air). Using the converged value of V_{bubble}, the corrected distance, X_2, can be calculated.

Visual comparison of TDR and physically measured distances in Figure 9-3 indicates increasing error in the TDR measurement with increasing transmission distance when the water level fell at unusually rapid rates (15 m/day). The error was smaller when water level dropped at a slower rate (2 m/day). This difference may result from agglomeration of water droplets in the tube during drawdown. The error should be very small at typical field rates of drawdown.

The TDR measurement overestimated (+2%) transmission distance during a rise in water level and an underestimated (−3%) transmission distance during a fall in water level. Correction for changes in V_p, resulting from water droplet effects, can be made to reduce this error to approximately 1% during periods of decreasing water level. The cable tester itself is accurate to only ±2% of the transmission distance, but this can be overcome by placing reference crimps in coaxial cable as discussed below. Other factors (e.g., temperature, cable degradation, deposition of minerals, etc.) that could alter V_p for a cable have not been investigated at this time.

When interpreting the TDR waveform, an error can be introduced if the operator does not properly interpret the TDR reflection at the air-water interface. Linear regression and tangent techniques such as shown in Figure 10-2 are widely used by soil science researchers and can be employed to minimize errors (Baker and Allmaras, 1990; Heimovaara and Bouten, 1990).

Coaxial Cable Offers Advantages Over Twisted Pair Wire

Dowding, Huang, and McComb (1996) found that there is less attenuation of the TDR voltage pulse and reflections with an air-dielectric coaxial cable (Cablewave SLA 38-50) than with twisted pair wire inserted into tubes. In addition, air-dielectric coaxial cable with a solid metal outer conductor allows use of crimps as distance reference marks, as mentioned in connection with deformation monitoring in Chapters 5, 6, and 7. Furthermore, this cable can be perforated with drill holes which not only allow free access of water but also produce reference TDR reflections.

The effect of crimping and perforating air-dielectric coaxial

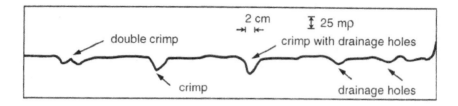

Figure 9-4. TDR reflections from crimps and drainage holes along air-dielectric coaxial cable (from Dowding, Huang, and McComb, 1996).

cable is shown in Figure 9-4. A 12 mm diameter cable (Cablewave SLA 12-50) was modified by crimping and by drilling drainage holes in the outer conductor. Six 4.7 mm diameter holes were drilled in a radial band (around circumference of outer conductor) on either side of a crimp. In addition, two more bands of drainage holes were drilled to the right of the crimp. The reduction of outer conductor metal in each band of holes was 56%.

Crimping and drilling drainage holes produced a larger amplitude reflection and facilitated migration of water around the crimp which, otherwise, would restrict water flow through the cable's annular space. As shown in Figure 9-4, the middle crimp (with drainage holes on either side) produced the largest magnitude reflection (approximately 28 mρ), followed by crimping alone (25 mρ), and the circumferential bands of holes (12.5 mρ). The magnitude of cable deformation was constant for both crimps.

To study the effect of crimps and holes on longer cables, a 12.7 mm crimp and twelve 2.4 mm diameter holes were added to the end of a 29.87 m long, 9 mm diameter coaxial cable (Cablewave SLA 38-50). Figure 9-5 shows waveforms from three different water levels in the vicinity of the crimp. The successive waveforms were acquired as the water level was raised in 15 mm increments. The smallest resolution of water level movement was 7.6 mm over a distance of 29.87 m or 0.025%. The magnitude of the crimp spike was 21 mρ. Holes were produced in an evenly spaced, radial band immediately adjacent to the crimp with six holes per side, which produced a 33% loss of metal in the outer conductor in the radial band. Thus, the crimp/hole combination at distances

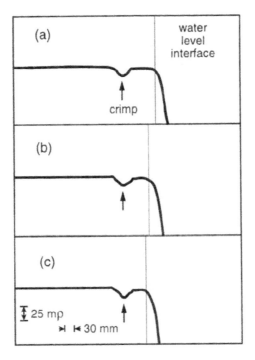

Figure 9-5. Change in TDR reflection as air–water interface approaches a reference crimp (from Dowding, Huang, and McComb, 1996).

associated with field measurements provides an effective distance marker.

Addition of crimp/hole combinations at selected locations is more useful for longer cables since the air–water interface can be located by distance from the reference crimp, rather than distance from the cable tester. Furthermore, the holes prevent formation of air pockets within the annular space between the inner and outer conductor. For cables longer than 100 m, a wider crimp or groups of crimps may have to be used to make a larger magnitude reference reflection.

The impedance mismatch at an air-water interface is so large that it can be measured using a half-sinewave TDR pulse. Compared with the step pulse used to acquire the waveforms in Figure 9-1, a half-sinewave pulse has the distinct advantage of reduced attenuation with distance by virtue of the lower frequency content of the pulse. Also, the pulse amplitude launched by commercially available half-sinewave TDR units is one or two orders of magnitude greater than the pulse amplitude launched by step-pulse TDR units (see Table 10.2). Thus, air-water interfaces can be detected along cables which are thousands of meters in length. This advantage will be illustrated later in this chapter with an example in which it was demonstrated that an existing piezometer at a U. S. Army Corps of Engineers dam could be retrofitted with air-dielectric cable and monitored using TDR.

APPLICATIONS AND CASE HISTORIES

Standpipe Piezometer

Geotechnical engineers traditionally measure ground water pressure with piezometers. Terzaghi and Peck (1967) review typical piezometers, including open standpipe, Casagrande, hydraulic, electrical, and pneumatic piezometers. Many of these systems are based upon equilibration of standpipe water pressure with that at the measurement point. The measurement point usually consists of a porous sand-filled screen or metal cylinder attached to a plastic standpipe, as shown in Figure 9-2.

While the standpipe equilibration technique is unusually robust, transducers used to remotely monitor water level typically depend upon "down-the-hole" electronics. When these electronics fail, the entire system is down. Reliable, manual measurement can be done using a pair of wires, uninsulated at their lower ends, which are lowered into the standpipe until the circuit is completed when the wires contact the water. TDR monitoring of air-dielectric coaxial cable makes it possible to combine the reliability of this manual technique with remote monitoring of "out-of-the-hole" electronics.

The height of water above the porous stone, shown in Figure 9-2, times the unit weight of water is equal to the water pressure at the elevation of the porous stone,

$$u_w = \gamma_w (D_s - X_2) \qquad\qquad (9\text{-}3)$$

where u_w is water pressure (kPa), γ_w is the unit weight of water (9.8 kN/m^3), D_s is depth to the porous stone (m), and X_2 is computed using Equation 9-1.

The use of TDR to measure piezometric levels has a number of advantages over other methods. Unlike down-the-hole pressure transducers placed at the porous stone, TDR electronics remain accessible above ground and can be protected against lightning strikes. In addition, one TDR cable tester can be employed to monitor several piezometers through multiplexing, as described in Chapter 10.

Retrofit of Dam Piezometer

The large voltage reflection at the air-water interface allows measurement at very large distances and the following example (O'Connor, 1996) demonstrates this possibility. The U. S. Army Corps of Engineers owns several dams at which piezometers are used to monitor pore water pressures. Many of these dams are instrumented with an automated monitoring system in which down-hole transducers are connected to a central processor via an extensive network of multipair lead wires. At one of these dams, a retrofit demonstration was conducted in which an air-dielectric cable was placed in one of the piezometer riser pipes (P-9 near Manhole No. 3 in Figure 9-6).

This demonstration showed that it is possible to retrofit dam piezometers with TDR technology using in-place multipair lead wire. In fact, this demonstration involved two different lead wires: 25-pair wire from the terminal box to Manhole No. 1 near piezometer P-6, and 2-pair wire from MH No. 1 to piezometer P-9. The transmission distance involved more than 600 m of lead wire, which required use of a high voltage (5 V compared with 0.3 V), half-sinewave TDR pulse rather than a step pulse TDR tester (see Chapter 10). A Tektronix 1503C cable tester was attached to the

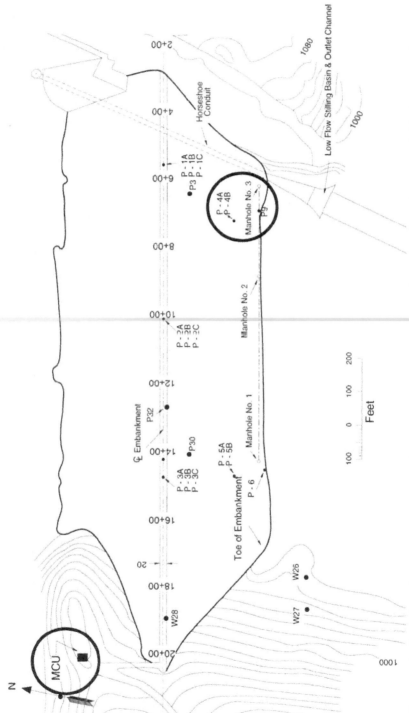

Figure 9-6. Plan view of dam, piezometer P-9 location, and readout location (MCU) (from O'Connor, 1996; Johnson, 1996).

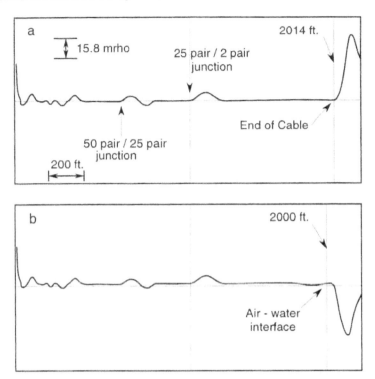

Figure 9-7. TDR waveforms acquired from lead wires and air-dielectric cable: (a) air-dielectric cable on ground; (b) air-dielectric cable in piezometer (from O'Connor, 1996).

25-pair lead wire at the terminal box at the USGS gage house on the right abutment (indicated as MCU in Figure 9-6). At the top of the piezometer riser pipe, the air-dielectric coaxial cable was connected to the 2-pair lead wire.

TDR signatures are compared in Figure 9-7 for the air-dielectric cable a) out of the standpipe, and b) in the standpipe. The air-water interface is clearly evident by the expected large amplitude reflection. The physical references used for this demonstration were the top and bottom of the air-dielectric cable. Water level was measured using TDR as follows:

TDR distance to bottom of cable	2014 ft
TDR distance to top of cable	<u>1987 ft</u>
TDR length of cable	27 ft

physical length of cable = 17 ft
calibration factor = 17/27

TDR distance to bottom of cable	2014 ft
TDR distance to air-water interface	<u>2000 ft</u>
TDR distance from interface to bottom	14 ft

TDR distance from interface to bottom of cable
= 14 ft (17/27) = 9 ft

This was verified manually using a two-wire water level indicator which measured the interface at 9.05 ft above the end of the cable. The reliability of the TDR measurement could have been increased by acquiring a baseline waveform before the air-dielectric cable was connected to the 2-pair lead wire to provide a reference at the end of the lead wire.

Water Levels Above and Within an Abandoned Mine

When cables are installed over abandoned mines to warn of subsidence, they are not intended to provide information about water levels within the abandoned mine or within the strata overlying the mine. However, water penetration into the foam dielectric coaxial cables has occurred often and it is considered worthwhile to discuss the value of the information obtained during the project described in connection with Figure 6-9.

Installation involved drilling holes from the ground surface down into the mine and placing a coaxial cable in each hole. A tube was then placed in the hole and grout, consisting of Portland cement, water, and an expansion agent, was tremied through this tube. The grout tube was then removed. The ends of the coaxial cables were not sealed prior to installation and this allowed water to penetrate the polyethylene dielectric at the cable end. As shown by the schematic in Figure 9-8, a variety of pathways could exist for water to flow from the rockmass into the cable. It is most likely that the cable is in contact with the borehole wall so that fracturing of the grout would not be required to allow water penetration.

Water penetration was apparent by the large reflection near

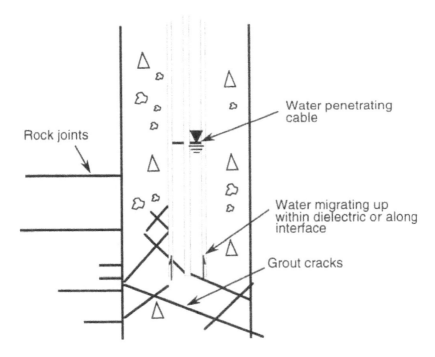

Rock joints

Water penetrating cable

Water migrating up within dielectric or along interface

Grout cracks

Figure 9-8. Water penetration in foam-dielectric coaxial cable installed in borehole above an abandoned mine; see cable TDR3 in Figure 6-9. Cable is most likely in contact with borehole wall.

the bottom of cables TDR2 and TDR3. The depth to this reflection for cable TDR2 is plotted versus time in Figure 9-9. By April 1996, the water inside both cables had essentially equilibrated at levels shown in Figure 6-9. The water level in TDR3 is consistent with that measured in the mine and the water level in TDR2 is consistent with the fractured shale stratum. There are clay layers at the top and bottom of this stratum which would allow it be a confined aquifer.

The height and rate of water level rise in the two cables appeared to be independent, implying that there was not communication between the shale aquifer and the mine. The water level rose asymptotically in both cables until equilibrium was reached as shown for TDR2 in Figure 9-9. The secant values of penetration rate for both cables are listed in Table 9.1 to illustrate how the rate decreased as the difference between water levels inside and outside the cable decreased (i.e., the hydraulic gradient decreased).

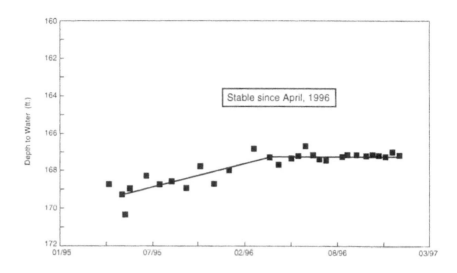

Figure 9-9. Time history of water penetration within foam dielectric cable; the cable was installed over an abandoned mine in hole TDR3 shown in Figure 6-9.

Table 9.1. Water penetration into grouted coaxial cable.

Hole	Date	Depth to water in cable, feet	Rate of water penetration	
			ft/day	10^{-6} cm/sec
TDR2	4/1/95	138.38		
	7/29/95	134.02	0.04	13.0
	5/17/96	128.84	0.02	6.0
	1/5/97	128.77	0.0003	0.1
TDR3	5/17/95	169.23		
	4/1/96	167.29	0.01	2.0
	1/5/97	167.19	0.0004	0.1

Water Level While Drilling

Accurate water table detection during drilling is an important aspect
of many environmental investigations. Such determination is
difficult in low permeability units, even in drilling operations where
a dry drilling fluid is used to overcome this challenge. A modified
two-wire TDR probe was designed by Hokett et al. (1994) to fit
within the drill rig's discharge line (Figure 9-10). The probe was
constructed from two cylindrical stainless steel rings (61 by 1.3 cm)
and designed to fit inside a plastic cylinder. A 5 cm break was cut
in each band to form the beginning and ending point of the waveguide. Separation of the two bands within the plastic cylinder was
maintained by a 5 cm plastic cylindrical spacer of the same thickness
as the metal ring. This design left one side of the stainless steel
rings exposed to the material flowing through the discharge line and
the other side against the plastic, minimizing probe abrasion. Given
this sensor geometry, the measurements were influenced by both the
plastic and the material flowing through the discharge line (Baker
and Lascano, 1989; Knight, 1992; Selker et al., 1993).

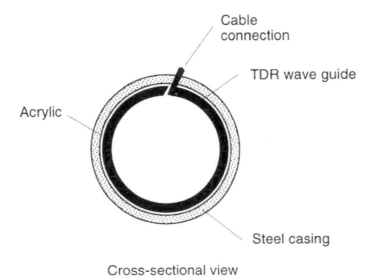

Cross-sectional view

**Figure 9-10. Schematic of waveguide installed in discharge
line of well-drilling circulation system (from Hokett et al.,
1994).**

A 1502B TDR cable tester interfaced with a 386 laptop computer was used to automatically measure and record the moisture content of the discharged drilling fluid. Custom software was developed that recorded, stored, and interpreted the TDR data. Measurements were normally recorded every 2 minutes while drilling was in progress.

The technique was tested on four U. S. Department of Energy wells drilled at the Nevada Test Site. In three of these tests, increases in the moisture content of the discharged drilling fluid measured by TDR coincided with the penetration of the water table. The data indicate that TDR may be useful for real-time field detection of the water table during drilling, but other drilling parameters can mask the TDR response.

Visual inspection of the drilling fluids during drilling of one well indicated water penetration at a depth of 530 m. Upon completion of drilling, the final water level in the well rose to 490 m, as expected. In this test, additional instrumentation was installed to measure the soap pump rate and compressor pressure. A direct relationship between pump stroke rate and raw TDR data was observed. This was evidenced by a pronounced shift in the TDR response (i.e., lower water content) between depths of 470 m and 490 m when pump strokes were reduced. However, additional slowing of pump strokes below 490 m did not produce a reduction in the TDR response. Hokett et al. concluded: 1) it was reasonable to assume that departure of the pump stroke trend from the TDR response was related to a contribution of water from the borehole below the water table, and 2) without pump stroke data, the TDR data alone in this test would not have been a useful indicator of water table penetration.

COMPARISON WITH OTHER TECHNOLOGY

While the standpipe measurement technique shown in Figure 9-2 is robust, electronic systems that remotely measure water pressure at the tip depend upon down-the-hole pressure sensors which are a weak link. When these down-the-hole electronics fail, the capability for remote operation is lost. There have been three problem areas associated with pressure transducers for which an abundance of field experience exists. First, transducers are susceptible to electrical transients such as lightning strikes. Second, the most reliable devices will not fit down into many of the smaller diameter riser pipes. Third, manual verification (field calibration) is difficult, particularly with the small diameter riser pipes. Any riser pipe smaller than 25 mm inner diameter presents both installation and calibration difficulties for commercially available pressure transducers.

TDR monitoring provides an effective solution to these challenges. First, sensitive TDR electronics are located above ground, where proven effective transient protection strategies can be implemented. The sensor or transducer is the cable, which is totally passive and not susceptible to transients commonly induced by lightning or power outages. Finally, the TDR measurement technique accommodates inherent self-calibration. The calibration reference is tied to the dimensional stability of the transmission line. Every measurement can be referenced to crimps made in a cable (Figures 9-4 and 9-5) at known locations, which eliminates the need for field verification.

Perhaps the most intriguing capability of TDR cable measurement is retrofitting existing piezometers. The summary of typical diameters presented in Table 9.2 shows the ranges of diameters as given by regulatory agencies (EPA) and standards organizations (ASTM), commercial products such as SINCO and ELE, and professional technicians and engineers including the U.S. Army Corps of Engineers. It is apparent that the minimum inside diameters for riser pipes in observation wells are 32 mm (1.25 in.), while those for piezometers are 9 to 12 mm (3/8 to 1/2 in.). Typical inside diameters are 38 to 51 mm (1-1/2 to 2 in.) for wells and 19 to 32 mm (3/4 to 1-1/4 in.) for piezometers. Typical transducer diameters are 9 to 27 mm (3/8 to 1 in.). The smallest diameter riser tube of 7 mm (1/4 in.) was associated with access to existing wells through ports on pumps.

Commercially available air dielectric coaxial cable can fit within the standpipes of most of the open system piezometers encountered in the field (Table 9.2), perhaps even including the

Table 9.2. Summary of typical wells, transducers, and piezometers diameters (Dowding et al., 1996).

Category	Reference	Diameter mm (inch)	
	Literature on Wells and Piezometers		
Observation well	EPA (1977)	32 to 914 (1-1/4 to 36)	ID
	ASTM (1990)	50 (2) minimum	ID
(well cluster)	EPA (1977)	51 to 63 (2 to 2-1/2)	ID
(multiple sample point)	EPA (1977)	ex: 32 to 102 (1-1/4 to 4), manufacturer dependent	ID
(existing pump)	EPA (1977)	7 (1/4) minimum	ID
Open standpipe	Mine Monitoring Manual (1990)	20 (7/8) minimum	ID
piezometer	EPA (1977)	schedule 40 PVC	ID
	Commercial Catalog Description of Transducers		
Slope Indicator Co.	Electro/Piezo system transducers	32 (1-1/4)	OD
(SINCO)	Hydro-pneumatic transducers	16 (5/8); 27 (1)	OD
	Water Level Indicator (manual)	9 (3/8)	OD
Soiltest of ELE Int.	Bailers	42.4 (1-5/8)	OD
	Oil/Water Interface Probe	25.4 (1)	OD
	Water Level Indicators	12.7 (1/2)	OD

Table 9.2. (continued)

Category	Reference	Diameter mm (inch)	
	Professional Technicians and Engineers		
Observation wells	STS Consultants, Illinois	38 (1-1/2) rare, 51 (2) common	ID
Monitoring well	U.S. Geological Survey (USGS)	38 (1-1/2) occasional, usually wells taper down to 51 (2)	ID
Casagrande piezometer	Geomation, Colorado	9 (3/8) smallest, 12.7 (1/2) occasional 19 to 32 (3/4 to 1-1/4) typical	ID
Open piezometer	U.S. Army Corps of Engineers	51 (2) PVC	ID

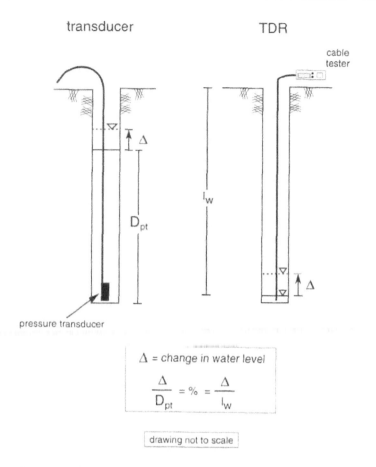

Figure 9-11.   Schematic of parameters used to compare resolution of down-the-hole pressure transducers and TDR measurement of water level.

smallest, which are 9.5 mm (3/8 inch) inner diameter. Air-dielectric coaxial cables cost from $3 to $12 per meter (1997 prices) and, therefore, result in relatively low costs for retrofitting existing piezometric systems.

It is important to compare the expected resolution of traditional pressure transducers and TDR monitoring of cable. Figure 9-11 shows a method for comparison of these approaches. An ideal comparison is one that challenges each system equally; the smallest water level movement (Δ) detectable at a long distance (pulse travel length to water level interface, l_w) for the TDR system and the smallest pressure change (head change in terms of height of water, Δ) measurable at a large pressure head in height of water (D_{pt}) for the transducer system. Resolution can be computed by dividing the small water level movement or head change (Δ) by either the pulse travel distance (l_w) or the pressure head in height of water (D_{pt}) for the respective systems. These ratios allow for comparison of resolution of these systems under equally challenging conditions.

Table 9.3 compares TDR resolution with that stated for commercially available transducers (Slope Indicator Co.). Resolution is defined as: a) the least digit the pressure transducer system is capable of displaying, or b) the smallest distance that two different air-water interface locations (waveforms) may still be distinguished or resolved using TDR.

The first transducer shown is the Vibrating Strip (V/S) Piezometer and VS DataMate. Product literature indicates a resolution of 3.5 mm of water with an accuracy of 0.1% of full-scale for the "standard" model, or 0.05% of full-scale for the "select" model. The comparison employed the mid-range (345 kPa or 50 psi range) for both models as it was able to sense the maximum pressure employed in the comparison. The second transducer is the Electro/Piezo crystal system with the model 56442 transducer and the model 56449 indicator. The accuracy is specified at 0.5% of range (after sensitivity and offset correction) with a resolution at 0.01% of the range. Again, the 345 kPa range was chosen for comparison to match the TDR cable lengths and possible pressure. The third transducer is the hydro-pneumatic system, models 514177 and 514178. The pressure range is 3.44 to 6,894 kPa (0.5 to 1,000 psig). The repeatability is given as ±0.346 kPa, or ±35 mm of water. The accuracy is given as a calibration zero offset of −2.0 kPa ± .35 kPa, or −210 mm ± 35 mm of water, and sensitivity of 1.007 ± 0.0005, where sensitivity is the ratio of output over input. The comparison was made assuming that the 210 mm of water head difference was evident at a pressure of 6,894 kPa (or 703 m).

The first TDR cable selected for comparison is a 27.4 m (90 ft) twin lead 300-ohm TV cable (Belden 8230). The second TDR cable is a 89.3 m (293 ft) twin lead 300-ohm TV cable (Belden

Table 9.3. Comparison of TDR and transducer resolution (from Dowding et al., 1996).

Measurement system	Length D_{pt} or l_w (m)	Change Δ[1] (mm)	Change Δ/D_{pt} (%)	Change Δ[1] (mm)	Change Δ/l_w (%)
		Transducer			
Short: V/S Standard	25	25	0.10		
Select	25	12	0.05		
Electric/Piezo	25	12	0.05		
Hydro-pneumatic	25	37	0.15		
Long: V/S Standard	87	87	0.10		
Select	87	43	0.05		
Electric/Piezo	87	43	0.05		
Hydro-pneumatic	87	35	0.04		
		TDR Cable			
TV, short:Prototype	24.6			24	0.09
Field	26.3			24	0.09
TV, long:Prototype	86.9			152	0.17
Field	87.8			91	0.10
Coaxial:Mod. Prototype[2]	29.6			8	0.02
Coaxial:Mod. Prototype[3]	83.6			< 19	< 0.025

[1] Δ = Smallest resolvable change in water level under given D_{pt} or lw condition shown in Figure 9-11.
[2] Only used air dielectric coaxial cable.
[3] Low loss coaxial lead cable + air dielectric coaxial transducer cable (see Figure 11-2).

8225). The third TDR cable is a 29.9 m (98 ft) air-dielectric coaxial cable (Cablewave Systems SLA 38-50). In comparison with the three commercial pressure transducers, the TDR results appear quite favorable as shown in Table 9.3, especially for the air-dielectric coaxial cable. It appears that the TDR system is capable of matching the resolution of pressure transducers while allowing the electronics to be kept out of the hole.

When considering field configurations, TDR cable design is quite flexible. For instance, the TDR cable can be divided into two portions: 1) lead wire portion and 2) sensing cable portion, as demonstrated in the dam retrofit example. These two portions may or may not be the same type of cable. TV cable, twisted pair wire, air-dielectric, or low-loss coaxial cable may be employed as transmission wire that leads to the air-dielectric coaxial sensing portion. The most cost-effective configuration appears to be a combination of braided copper low-loss lead cable ($0.50/ft, 1997) and air-dielectric transducer cable ($4.00/ft, 1993), as shown in Figure 11-2.

DETECTION OF LEAKING LIQUIDS

PermAlert has developed the system shown in Figure 9-12 in which TDR is used to monitor leaks along sensor strings. The TDR reflections are specific to the condition of the installed sensor cable and are stored in memory as a reference map. An alarm unit continuously updates the TDR reflections and compares them with values of the benchmark TDR waveform stored in memory. When a leak occurs, the liquid wets the sensor cable which alters the cable's impedance at the leak location. The monitoring unit's microprocessor recognizes the TDR reflection from the wet portion of cable and activates an alarm. A new reference map with the change can be stored in memory to allow monitoring to continue.

The system locates the point of origin of a leak or cable fault within ±1% of the distance from the last calibration point or ±5 feet, whichever is greater. In alarm mode, the unit activates output relays to facilitate the control of valves or remote alarms (if

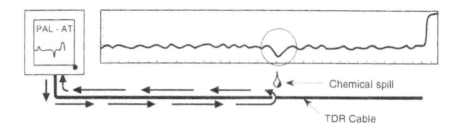

Figure 9-12. Detection and location of fluids penetrating specially designed coaxial cable using TDR (from PermAlert, 1995).

desired), while providing audio and visual alarms, including a digital display of the distance to the leak origin.

Sensor Cable

The sensor cable shown in Figure 9-13 includes a central electrical conductor, annular spacer of electrically insulating material around the central conductor, and an outer sheath (Bailey, 1994). The outer sheath spacer is formed of a braid of electrically conductive wires, which are coated with corrosion resistant material for protection against water, acids, alkalis, solvents or other liquids, or environmental contaminants that may be present in the vicinity. The annular spacer has voids into which leakage fluid can enter.

Barlo et al. (1996) have suggested that distance markers produced by permanent geometrical or property changes in such cable would provide reference reflections at known locations for comparison with the voltage reflections produced by changes in the dielectric constant when contaminant liquids penetrate. Their precise location would improve the location accuracy.

Sensor String Components

A sensing string can be made up of any combination of probes and sensor cables. Sensor cables and probes are designed to serve a wide variety of applications. The system is not subject to false alarms caused by dust or other nonliquid conductive materials that

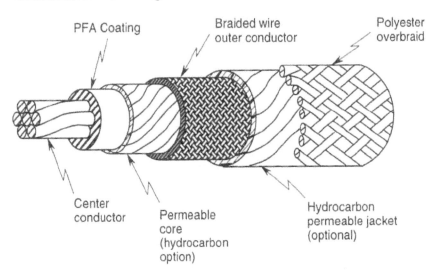

Figure 9-13. Construction of coaxial cable designed to allow detection of hydrocarbons below the water table (from Bailey, 1994. U.S. Patent 5,355,720).

may come into contact with the cable, or from casually stepping on the cable. Several sensor cables (Table 9.4) are capable of detecting and locating both water-based and hydrocarbon liquids, while others will detect only hydrocarbons, ignoring water. In most applications, the sensor cable can be dried and reused after a leak is repaired and clean-up has been completed. State-of-the-art "gold cables" have no exposed metal and are designed for corrosive chemical applications. Sensitivity is the length of sensor cable that must be wetted with a specific liquid before an alarm condition occurs. Sensitivity is a function of sensor string length and monitoring panel settings. In addition to sensor cable, PermAlert provides jumper cables. Jumper cable is used to connect the monitoring unit and sensing string or link sensing strings between monitored areas.

Probes monitor for leaks at specific locations. There are several probes (Table 9.5) available to monitor for water and/or hydrocarbons that can be connected in series to the sensing string for a wide variety of applications.

Table 9.4. Sensing cables (from PermAlert, 1995).

Cable	Action	Sensitivity/Resistance	Applications
AGW- Gold	quick drying cable	detects highly corrosive liquid leaks such as acids, bases, and solvents; chemically resistant	typical applications are secondary contained pipes, subfloors of clean rooms, and high temperature applications
AGT	wicking cable requires more drying time than AGW-Gold	detection of accumulations at a shallow depth of 2 mm (1/16 in.) of liquid; corrosion resistant for most applications	can be strapped to the above-ground single-wall pipes and used under or around equipment applications in subfloor or other flat surface leak monitoring applications
AGT-Gold	wicking cable	detects highly corrosive acid, base, and solvent leaks; chemically resistant	clean room subfloors, above ground single wall pipes, and equipment applications
TFH	wicking cable	detects only hydrocarbons.	this cable may be direct buried to a maximum depth of 20 feet to locate fuel leaks while ignoring the presence of water
TFH-Gold	wicking cable	detects only hydrocarbons; corrosive applications	applications include locations where hydrogen sulfide, or other corrosive gases or liquids, may be encountered

Table 9.5. Probe sensors (from PermAlert, 1995).

Probe	Action	Sensitivity	Applications
PHL		will detect only hydrocarbon liquid	typically installed in the interstitial space of double-wall tanks
PWS		will detect water-based liquids	used in tandem with PHL probes to monitor double-wall tanks containing hydrocarbon products for water and hydrocarbon leak detection
PSTV	float switches		designed for installation in a 50 mm (2-in.) standpipe on double wall tanks. It allows monitoring of the interstitial space of a double-wall tank or high/low level of the product in the tank
PTHL	float switches		designed for installation in a 50 mm (2-in.) standpipe on double wall tanks. It allows monitoring of the interstitial space of a double-wall tank or high/low level of the product in the tank
PHFW		designed to monitor hydrocarbons floating on water	installs in a 100 mm (4-in.) diameter monitoring well and can be used for ground water or sump monitoring
PFS	float switches	will detect water-based or hydrocarbon liquids	designed to monitor liquids in manholes, sumps, etc.

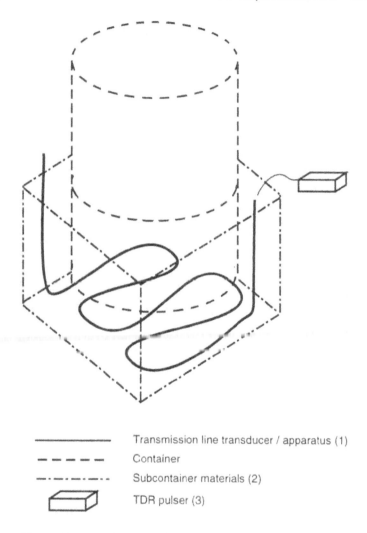

─────────────── Transmission line transducer / apparatus (1)

─ ─ ─ ─ ─ Container

─··─··─·· Subcontainer materials (2)

TDR pulser (3)

Figure 9-14. Schematic of possible layout for detection of leaks beneath a storage tank.

Application of Leak Detection

Figure 9-14 is a schematic diagram of the leak detection system consisting of a length of the sensing cable and TDR pulser/sampler. The cable may be buried or placed in slotted conduit. The cable may be placed in any geometry that optimizes detection and it can be pulsed from both ends to maximize location accuracy.

One application would be monitoring of underground storage tanks to detect leaks and allow sufficient time for the tanks to be emptied before leakage continues. The Illinois Legislature recently passed a law authorizing a surcharge on the delivery and sale of petroleum products by licensed distributors (Tarnoff, 1996). The surcharge, expected to generate about $60 million per year, will replenish a state fund established to reimburse owners and operators of leaking underground storage tanks for cleanup associated with tank removal and upgrades. During the summer of 1996, the program was taken over by the federal Environmental Protection Agency (EPA), which requires that owners and operators purchase $1 million in insurance coverage for the sites.

Passage of the law brought some relief to tank owners and operators such as gas stations, hospitals, and school districts that were uncertain whether they would be reimbursed for cleaning up environmental damage so they could meet the 1998 federal EPA deadline. The law imposes a $60 surcharge for every 30,000 liters (7,500 gallons) of petroleum products delivered and sold by a licensed distributor. It is intended to replenish a state fund created by the Illinois Underground Storage Tank (UST) program that reimburses owners and operators of tanks who clean up contaminated soil associated with the upgrade, removal, or replacement of tanks after a $10,000 deductible. In some cases, parent oil companies pay for the clean up.   Nevertheless, tank removal causes a lot of inconvenience and loss of revenue. The owner of one station estimated that the station lost $160,000 in gross gasoline sales during a three-week period.

There are more than 49,000 active underground storage tanks in Illinois (Tarnoff, 1996). In addition, many tank problems go unreported because many small businesses don't have the money to clean them up. Barlo et al. (1996) estimated that there are more than 1,000,000 active USTs in the U.S.

Chapter 10
ELECTRONICS

While various components of TDR systems can be acquired from a variety of suppliers, the basic relationship shown in Figure 10-1 remains constant. Since the state of the art in instrumentation is dynamic and is expected to be constantly changing, components of a TDR system will be discussed in a generic manner. The discussion begins with the sensor/transducer components, followed by connections from the sensors to the TDR pulser/sampler. Next, system control methods are discussed, followed by components for storage and downloading of TDR data. Finally, power requirements for remote monitoring are addressed and detailed examples of systems which have been used to monitor rock deformation and soil moisture are summarized.

MOISTURE PROBES AND TRANSMISSION LINE TRANS-FORMERS

A summary of commercially available moisture probes is given in Table 3.2. Typically, the probes consist of two or three parallel stainless steel rods that are 2 to 3 mm in diameter, 15 to 50 cm long, and spaced 25 to 30 mm apart. The ratio of diameter to spacing, d/s, is greater than 0.10. As shown in Figures 2-7, 2-10, and 3-1, characteristics of the TDR waveform depend on the probe design and conditions within the porous medium in which a probe is embedded.

Early in the development of these probes, Stein and Kane (1983) used two transformers to bridge 50 ohm coaxial cable to 300 ohm two-wire TV cable as shown in Figures 10-2 and 10-3a. This complex consisted of a) 50-75 ohm transformer, b) 75 ohm cable, c) 75-300 ohm TV transformer, and d) 300 ohm TV antenna cable that connected to the two rods with threaded connectors. The two-rod probe and a portion of the 300 ohm cable were buried in the soil.

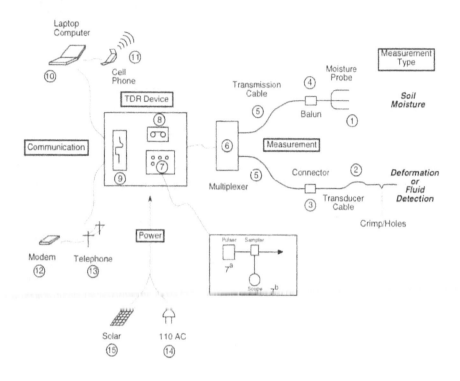

Figure 10-1. Components of TDR monitoring systems; numbers indicate the order in which components are discussed in this chapter.

As shown in Figure 10.2, determination of the beginning and end of the probe was made through use of tangents to the trace, as first reported by Patterson and Smith (1981). A profile view of the probes is shown in the upper part of the figure, starting with the TV line, followed by the nuts at D, and finally the probes. The transition zones on the trace were A-E and V-Z, where D was the beginning of the probes and W was the end of the probes. The trace went down at point A, because the impedance of the antenna cable buried in soil was lower than the impedance of the cable in air. Each transition zone differed according to the design used and the dielectric material between the probes. In the case of a 300 ohm line in the soil, point W lay between C and D most of the time

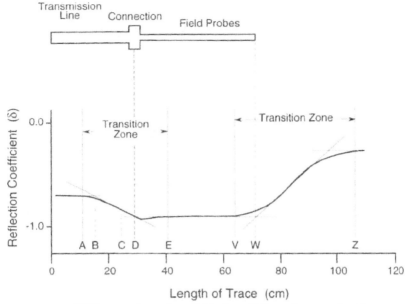

Figure 10-2. TDR waveform for a 300 ohm 17.7-cm long probe (from Stein and Kane, 1983).

because of the magnitude of power loss in the 300-ohm line. In such cases it was impossible to find the beginning of the probes (i.e., D) through the use of tangents because the curve did not flatten between B and W.

The next modification involved using 50-ohm coaxial lead cable, as shown in Figure 10-3a. A balun transformer was needed to connect the coaxial cable to the probe rods, and the transformer itself caused TDR reflections. This could be remedied by not having the transformer too close to the probe. Other balun design considerations are discussed later in this chapter.

The series of transformers required with a 2-rod probe was next simplified with the 3-rod probe shown in Figure 10-3b. The 50-ohm coaxial cable was connected directly to the probe rods which eliminated the transformer-induced reflection as shown by the waveform in Figure 10-3b. Further modifications have been made with respect to the diameter, spacing, and length of probe rods which have enhanced the 3-rod probe performance and durability (Table 3.2).

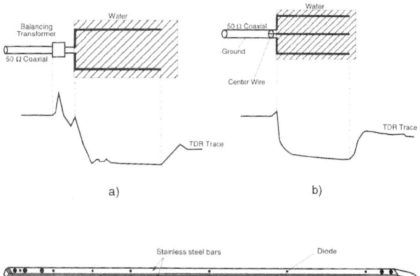

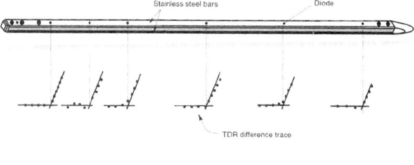

Figure 10-3.   2-Rod, 3-rod, and segmented-rod probes for soil moisture monitoring: (a) waveform obtained from 2-rod balanced probe; (b) waveform obtained from 3-rod unbalanced probe; (c) difference waveforms for segments of MoisturePoint™ probe (from Whalley, 1993; ESI Environmental Sensors, Inc., 1996).

Recently, ESI Environmental Sensors, Inc. has developed a technique to make measurements along a stiff 2-rod segmented probe. This device is shown in Figure 10-3c. The MoisturePoint™ probe is discretized into segments of various lengths by diodes which are switched on and off. By virtue of measuring the travel time along each segment, it is possible to obtain a profile of moisture content versus depth with a single probe that is driven into the ground. In general, the probe is a long rod with a rectangular cross section and a length determined by the number of segments and their aggregate length. The material construction is a molded sandwich, 13 mm x 19 mm x (probe length) of two thin stainless steel bars and epoxy. Epoxy is injected between the two bars to fix their separation distance so, physically, the probe looks like a hard, black spear with stainless steel sides. The segments are defined by diodes molded into the epoxy at selected locations along the probe. TDR waveforms are obtained with the diodes switched off and on and the difference waveforms are used to determine the travel time between diodes. This approach inherently eliminates the need for baluns to remove unwanted reflections, but this approach still has limitations.

PROPERTIES OF COAXIAL CABLES

The smooth-wall aluminum coaxial cable shown in Figure 10-4a has an annular space which is fixed and uniform in geometry. The coaxial cable shown in Figure 10-4b is constructed with a corrugated outer conductor for increased flexibility and easier bending as compared with the smooth-wall aluminum cable. Braided cables (Figure 10-4c) are even more flexible than the corrugated cable. Common cable diameters are 9.5 mm, 12.5 mm, 19 mm, and 22.2 mm (i.e., 3/8-in., 1/2-in., 3/4-in., and 7/8-in.). Air-dielectric cable (Figure 10-4d) is used, with some modification, for measuring water levels. Engineering and electrical properties for various coaxial cables are listed in Table 10.1 and discussed in Chapter 2. Vendors are listed in Appendix F.

TRANSDUCER CABLES

This discussion focuses on that portion of a cable used as a transducer, whether it be for deformation monitoring, fluid level monitoring, or other geoapplications of metallic TDR technology. As discussed in various chapters of this book, smooth-walled aluminum and corrugated-copper foam dielectric cable is used for

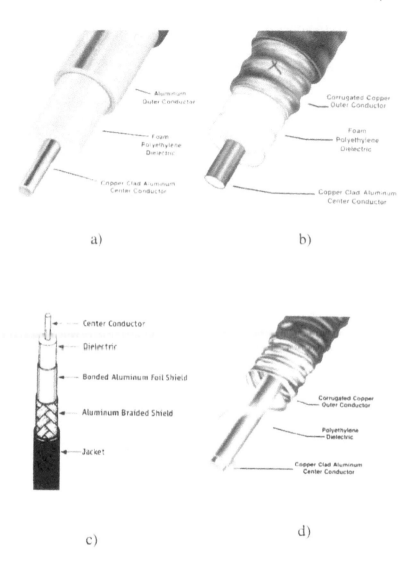

a)   b)

c)   d)

Figure 10-4. Construction of commonly used coaxial cables: (a) smooth-wall aluminum outer conductor (Cablewave Systems FXA12-50); (b) corrugated-copper outer conductor (Cablewave Systems FLC12-50); (c) braided-aluminum outer conductor (©1998CommScope, Inc. of North Carolina F1160BEF); (d) corrugated-copper air dielectric (Cablewave Systems HCC12-50).

Table 10.1a. Properties of commonly used smooth and corrugated coaxial cables.

Manufacturer	Vendor ID	RG/U type	Cost (1997) ($/m)	Inner conductor[1] OD (mm)	Dielectric type	K	Outer conductor ID (mm)	OD (mm)	nominal thickness (mm)	Min. bending radius	Max. pulling tension (N)	Weight (N/m)
Smooth aluminum outer conductor												
Cablewave	FXA12-50	RG331/U	$5.48	4.09	cellular polyethylene foam	1.5	11.43	12.70	1.27	127	1792	2
	FXA78-50	RG332/U	$16.24	7.31			20.34	22.22	1.88	254	4676	6
Comm/Scope	P-3 75-875CA	RG336/U	$2.36	4.93	expanded polyethylene	1.5	20.24	22.23	1.99	343	3500	5
	P-3 75-750CA		$1.97	4.29		1.5	17.22	19.05	1.83	254	2700	4
Corrugated copper outer conductor												
Cablewave	FCC 38-50			3.02	cellular polyethylene foam	1.5	7.57	9.53	1.96	100		2
	FLC 12-50		$8.20	4.83		1.5	11.43	13.72	2.29	127		4

[1] copper-clad aluminum

Table 10.1b. Properties of commonly used air-dielectric cables and braided cables.

Manufacturer	Vendor ID	RG/U type	Cost (1997) ($/m)	Inner conductor[1] type	OD (mm)	Dielectric type	K	Outer conductor[2] type	ID (mm)	OD (mm)	nominal thickness (mm)	Min. bending radius (mm)	Max. pulling tension (N)	Weight (N/m)
						Air dielectric								
Cablewave	SLA 38-50		$11.64	A	3.40	air		D	8.26	9.53	1.27	102		1
Cablewave	HCC12-50		$8.36	A	4.06	air		E	8.97	12.29	3.32	127		1
						Braided outer conductor								
CommScope	F1160BEF	RG11/U	$0.46	B	1.63	expanded polyethylene	1.5	F	7.11	8.96	1.85	75	1460	1
Belden	8214	RG8/U		C	2.31	polyethylene	2.3	G	7.24	10.29	3.05			2

[1] A = copper clad aluminum   [1] B = copper clad steel   [1] C = twisted copper
[2] D = solid aluminum   [2] E = corrugated copper   [2] F = braided aluminum   [2] G = braided copper

Table 10.1c. Electrical properties and propagation velocities of commonly used coaxial cables.

Manufacturer	Vendor ID	Distributed Capacitance (nF/km)	Characteristic Impedance (ohms)	Propagation Velocity	Series DC Resistance (ohms/km)		
					inner conductor	outer conductor	loop
Cablewave	FXA12-50	82.0	50	81%	12.79	11.81	
Cablewave	FXA78-50	80.4	50	81%	41.96	38.73	
Comm/Scope	P-3 75-875CA	50.0	75	87%	1.38	0.43	1.81
Comm/Scope	P-3 75-750CA	50.0	75	87%	1.87	0.62	2.49
Cablewave	FCC38-50		50	81%			
Cablewave	FLC12-50		50	88%			
Cablewave	SLA38-50		50	90%			
Cablewave	HCC12-50		50	92%			
Comm/Scope	F1160BEF	53.1	75	85%			
Belden	8214	85.3	50	78%	3.90	3.60	

Table 10.1d. Attenuation of commonly used coaxial cables.

Manufacturer	Vendor ID	Attenuation, (dB/100m)					
		5 MHZ	10 MHZ	30 MHZ	50 MHZ	400 MHZ
Cablewave	FXA12-50	0.59		1.56		6.56
Cablewave	FXA78-50	0.32		0.92		4.19
Comm/Scope	P-3 75-875CA	0.30		0.79	1.05	2.99
Comm/Scope	P-3 75-750CA	0.36		0.85	1.15	3.44
Cablewave	FCC38-50			1.93		8.20
Cablewave	FLC12-50			1.21		4.59
Cablewave	SLA38-50			1.96	6.95	6.95
Cablewave	HCC12-50			1.49		5.64
Comm/Scope	F1160BEF	1.25				8.53
Belden	8214		1.70	2.70	3.90	13.80

rock and soil deformation monitoring, corrugated-copper cable is used for structural deformation monitoring, and air-dielectric cable is used for water level monitoring.

While the primary objective in cable selection has been to minimize cost, other factors have overridden expenses. For example, experience has proven that employing inexpensive cable in poorly grouted holes can result in failure to detect shearing deformation. Attenuation of voltage signals and electrical noise degrade the signal-to-noise ratio of TDR reflections. For example, many recent (1990 to 1998) attempts to monitor and quantify rock deformation with relatively inexpensive braided copper antenna cable bound to a slope indicator casing have not been as successful as solid aluminum outer conductor cables grouted in their own holes. Appropriate cables must be grouted into their own holes to adequately monitor rock or soil deformation. In cases where the objective is to monitor fluid levels, it is possible to employ either air-dielectric coaxial cable or 2-wire TV antenna cable. The air-dielectric coaxial cable (Figure 10-4d) is preferable due to the reduced attenuation and higher signal-to-noise ratio.   When monitoring structural deformation, the outer corrugated conductor (Figure 10-4b) is more sensitive to extension than smooth-wall coaxial cable (Figure 10-4a).

The electrical properties that have the greatest impact on the quality of TDR reflection waveforms are 1) resistance and 2) attenuation. Both these factors control the magnitude of TDR reflection for a given magnitude of cable deformation. This is one reason that braided cable has not been extensively deployed as a transducer.

The mechanical cable properties that have the greatest impact on TDR monitoring of rock and soil deformation are 1) outer conductor (solid or braid), 2) minimum bending radius, 3) tensile strength, 4) diameter of outer conductor, and 5) dielectric type, which controls stiffness. A solid metal outer conductor not only makes it possible to create the most permanent reference crimps as shown in Figures 5-12 and 9-4, but also greatly enhances the sensitivity to shear deformation. Braided cables with a Teflon dielectric (Figure 10-4c) are resistant to deformation; however, it has been found that braided cable with air dielectric may provide the optimal response for monitoring localized shearing in soil as discussed in Chapter 7. The amount of movement that must occur before a TDR reflection can be detected and the reflection magnitude corresponding to a given deformation is controlled by cable diameter. Furthermore, cable diameter determines the range of movement that can be detected before a cable is short-circuited (Su, 1987).

In order to maximize the bond between the outer conductor and grout, cables should be ordered without an outer plastic jacket, if possible. In some situations, such as some *in situ* mines discussed in Chapter 6, the downhole environment is corrosive and a jacketed cable should be specified.

BALUNS OR TRANSMISSION LINE TRANSFORMERS

When using TDR to monitor soil moisture, the output from a pulser along a coaxial lead cable is unbalanced (i.e., the center conductor carries the signal while the shielding is grounded). If a 2-rod probe is used, as shown in Figure 10-3a, an impedance matching transformer is used to convert the unbalanced output from the TDR to a balanced output so that each of the rods carries a signal which is equal in magnitude but opposite in sign. Zegelin et al. (1989) bypassed the transformer by directly connecting the unbalanced output from a 50-ohm coaxial cable to an unbalanced 3-rod probe as shown in Figure 10-3b. In this system, the central rod is connected to the center conductor of the coaxial cable and the outer rods are connected to the shielding of the coaxial cable.

Baluns are a subset of broadband, impedance-matching transformers (Sevick, 1990), and serve two purposes: (a) to change an electrical field from balanced to unbalanced or vice versa, and (b) to match lines with different impedances. A balanced to unbalanced transformer can be constructed by simply winding a transmission line through the holes of a ferrite core. The impedance-matching characteristics of the transformer are specified as the ratio of the input and output impedances of the balun. The schematic for a 1:1 balun is given in Figure 10-5c. Matching lines with different impedances can be accomplished with a transformer first suggested

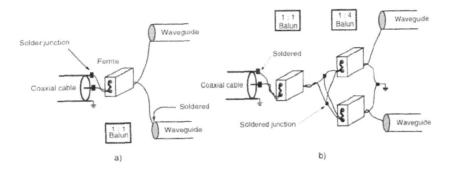

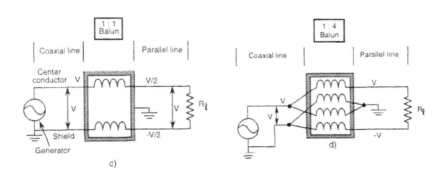

Figure 10-5. Baluns used with soil moisture probes: (a) 1:1 balun; (b) 1:1 balun and 1:4 balun;(c) schematic of 1:1 balun which converts the electrical field from unbalanced to balanced; (d) schematic of Guanella-type 1:4 balun showing continuous electrical path from central conductor to ground (from Spaans and Baker, 1993).

by Guanella (Sevick, 1990). Sevick stated that Guanella's proposed transformer features superior performance compared with other transformers. e.g., greater bandwidth and bidirectional capability. Figure 10.5d shows the schematic of a Guanella 1:4 balun, which consists of two 1:1 baluns in a series-parallel arrangement. A perfect 1:4 balun can match 50-ohm coaxial cable to 200-ohm parallel cable without causing any partial reflections.

One of the baluns that Spaans and Baker (1993) built and tested, the SB1:4 [designated S(oil) B(alun) 1:4], matches 50-ohm unbalanced to 200-ohm balanced lines as shown in Figure 10-5b. When they used TDR methods to measure electrical conductivity, results obtained with a SB1:4 balun and a conventional TP-103 balun matched poorly with measurements made with a conductance meter. They suggested that this divergence may have resulted from low-frequency signal attenuation in the 1:4 balun, which is more pronounced in the SB1:4 than in the TP-103, since the signal is shorted to ground through the transformer.

The objective was to use TDR to indirectly measure conductivity by quantifyiing attenuation of the low-frequency signal returned from the end of the probe, as shown in Figures 2-10, 2-11, and 2-13. Spaans and Baker developed an alternative SB1:1 balun shown in Figure 10-5a. Its performance in soil moisture content determination is at least as good as that of 1:4 baluns, and the asymptotic shape of the trace after the final reflection is appealing for TDR-based methods of conductivity determination. Spaans and Baker also found additional advantages of the SB1:1 were shorter fabricaton time and lower cost.

CONNECTORS

In cases where all transducer cables, probes, lead cables, and multiplexers are supplied by one manufacturer or distributor, it is possible that a consistent type of coaxial connector will be employed. Furthermore, if all cable lengths can be specified and connectors attached before shipping, then consistent connectors will be used. However, it is more likely that cables will have to be cut to length at a site requiring field assembly of connections.

One of the most common connectors used with RF instrumentation is the BNC type. However, there is a plethora of coaxial connectors available (F, N, UHF, etc.) as well as two genders. Therefore, an inventory of adapters will need to be maintained. Often, the type of probe, transducer cable, or lead cable will dictate the use of various connectors. Since these will be different, a variety of adaptors will also be required.

Connections are the weak link in a TDR monitoring system. Consequently, they require the greatest attention during design, installation, and maintenance. In cases where connectors will be buried, they should be placed in waterproof junction boxes that allow access for maintenance. In cases where connections will be assembled in the field, it is best to use a type which doesn't require soldering. This will really come back to haunt you. At junction boxes, an extra length of cable should be used not only for purposes of minimizing stress on connectors, but also to allow cutting of cable to assemble new connections.

Secure connections are critical since these inherently can cause impedance changes and attenuation of both the transmitted and reflected voltage. On a positive note, connections also cause TDR reflections at known physical locations so they provide distance calibration every time that waveforms are acquired.

LOW-LOSS LEAD CABLE

Lead cable is basically a long connection between the pulser/sampler and transducer cable or probe. It is critical to use shielded low-loss lead cable (e.g., Figure 10-4c) to reduce noise and attenuation. Ideally, the lead cable would be the same as the transducer cable, but cost can be prohibitive. The alternative is to use a less expensive cable (e.g., RG6, RG8, or RG11 listed in Table 10.1) that will cause little attenuation of the transmitted and reflected voltage over cable lengths of 100 m or more.

Lead cable will extend from the multiplexer to transducer locations laid out on the ground, buried in shallow trenches, or suspended from poles. If there are high-voltage power lines or radio transmitters within 300 m of the lead cables, it is very likely that electrical noise will be picked up by the pulser/sampler along the lead cable. This can be minimized by ensuring that the cables are shielded. The shield and pulser/sampler should be connected to a common ground to prevent ground loops.

MULTIPLEXER

If a system has multiple functions but is costly, its usefulness is reduced. If a system is economical but lacks many essential features, its usefulness is also reduced. Multiplexing of several cables or probes to one TDR pulser/sampler makes it possible to simultaneously monitor deformation, water level, soil moisture, conductivity, etc.

A variety of multiplexers are available. Baker and Allmaras (1990) describe a mechanical coaxial switch. Heimovaara and Bouten (1990) describe a 36-channel coaxial relay. One of the most commonly used RF multiplexers for TDR systems is the SDMX50 (Campbell Scientific, Inc., 1991) shown in Figure 10-6.

Multiplexed systems can be employed with multiple types of measurements in one monitoring program that requires recording of a wide variety of TDR waveforms. For example, assume that a project requires simultaneous monitoring of water level, rock/soil shearing, and soil moisture. The system shown in Figure 10-6 can be deployed to fulfill this requirement. A 100 m long low-loss cable could be attached to the first channel. At the end of this low-loss lead cable, a 10 to 20 m long air-dielectric coaxial cable could be employed as a probe to measure water level changes. A 500 m long coaxial cable could be connected to the second channel to measure rock/soil shearing. Similarly, 30 m lead cables and parallel-rod probes could be connected to the remaining channels to monitor soil moisture. Each channel would be used with different acquisition settings to be compatible with the variety of waveforms.

PULSER/SAMPLER

A summary of commercially available TDR pulser/samplers is listed in Table 10.2. Alternatively, an inexpensive TDR can be assembled for under $1,500 using a conventional 100 MHz oscilloscope and a general-purpose pulse generator (Andrews, 1994). Inexpensive pulse generators with rise times less than 5 ns are available. A 20

Figure 10-6. Multiplexed system used for remote monitoring of soil moisture and conductivity.

dB BNC attenuator should be attached to the pulse generator output to assure a good 50-ohm source impedance. A BNC tee adapter is connected directly to the scope 1 Megohm input. Attach the pulse generator and attenuator to one arm of the tee. The other arm is the TDR output port. Do not use any coaxial cable between the BNC tee and the scope as this would cause undesirable multiple reflections, unless it can be correctly terminated. If this pulser and attenuator are used with a 100 MHz sampling oscilloscope, the result is a TDR with a rise time of 6 ns and a spatial resolution of 60 cm in ordinary coaxial cable (Andrews, 1994).

The range of common TDR signal pulses is shown in Figure 10-7. A step pulse (Figure 10-7a) is frequently used as the TDR test signal, because the long plateau after the step conveys DC information about the reflecting object, while the very short rise time of a step pulse contains very high frequencies and thus gives HF information and good spatial resolution (Andrews, 1994).

Table 10.2. Commercially Available Metallic TDRs (after Andrews, 1994).

Manufacturer	Type	Pulse type	Pulse amplitude (V)	Dimensions (mm)	Weight (kg)	Power source	Range (km)
IWATSU	oscilloscope / sampling head	<45 ps step				AC	
LE-CROY	oscilloscope / sampling head	<150 ps step				AC	
HYPERLABS	TDR plug-in card	35 ps step				IBM PC	
	TDR	200 ps step	0.3	191 x 89 x 50	2	battery	10.0
Tektronix	oscilloscope / sampling head	<35 ps step				AC	
	TDR	200 ps step	0.3	127 x 315 x 436	9	battery	0.6
	TDR	½ sine wave (2, 10, 100, 1000 ns)	5.0	127 x 315 x 436	9	battery	15.0
	TDR	½ sine wave (20 to 3000 ns)	20.0	30 x 25 x 9	3	battery	15.0
Hewlett-Packard/	oscilloscope / sampling head pulser	<45 ps step				AC	
Picosecond Pulse Labs		18 ps step				AC	

Table 10.2. (continued)

Manufacturer	Type	Pulse type	Pulse amplitude (V)	Dimensions (mm)	Weight (kg)	Power source	Range (km)
ESI Environmental Sensors, Inc.	TDR	140 ps step	0.3	274 x 248 x 173	5	battery	0.1
Soilmoisture Equipment Co.	TDR	200 ps step	1.5			battery	0.1
Easy Test Ltd.	TDR	200 ps needle shape		260 x 180 x 130	4	battery	0.1
CM Technologies Corp.	TDR plug-in card	200 ps step (?)				IBM PC	
IMKO GmBH	TDR	200 ps step	1.0			battery	0.05
PermAlert ESP	TDR					AC	
Riser-Bond Instruments	TDR	½ sine wave (2 to 4000 ns)		267 x 248 x 127	4	battery	19.0
Biddle Instruments	TDR	½ sine wave (30 to 1500 ns)		140 x 290 x 260	8	battery	15.0
Signatel	TDR	200 ps step	0.6	266 x 241 x 114	3	battery	1.5
Tenzor	TDR plug-in card	100 ps step				IBM PC	0.1
Bicotest	TDR	½ sine wave (2 to 1200 ns)	2.5	75 x 183 x 300	3	battery	12.0

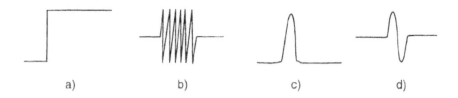

Figure 10-7. Common TDR test pulses: (a) step pulse; (b) radar pulse; (c)
half sinewave; (d) monocycle (from Andrews, 1994).

Other pulse types are used for different applications. The radar
pulse shown in Figure 10-7b is a short sinewave, which is turned on
briefly and is ideal for testing narrow-bandwidth systems, such as
waveguides. For systems that do not support DC, narrow half-
sinewave or monocyle impulses (Figure 10-7c, d) are often used.
The half-sinewave impulse is used with several commercially
available TDRs (see Table 10.2) with a major trade-off being
spatial resolution versus range.

The commercial, state of the art in coaxial cable TDRs is a
10 ps rise-time instrument made by Hewlett-Packard (HP) and
Picosecond Pulse Labs (PSPL) (Andrews, 1994). This consists of
the HP-54124A, 50 GHz, 9.4 ps rise time, digital sampling oscilloscope, and the PSPL 4015B, 9 volt, 15 ps pulse generator. The
result of combining this very fast sampler, pulser, and microprocessor is a powerful TDR instrument with a 10 ps rise time. This short
rise time provides a spatial resolution of 1.0 mm in coaxial cable.

Malicki (1990) described a time domain reflectometer which
produces a needle shaped pulse in contrast to the more common
step shaped pulse. The pulser/sampler is claimed to provide a
clearer definition of the start and end of the transmission line
embedded in soil than the more widely used step pulse.

The HYPERLABS HL1500 is an example of a growing
number of compact systems. It is small enough for packaging with
other instruments. The miniature pulsing card is linked to a
separate microprocessing card (sampler), forming the core of the
TDR shown in Figure 10-8. Laboratory tests show that the pulsing
card performs accurately in temperatures ranging from −40°C to
+85°C. Power consumption is designed for a maximum of 1 watt.

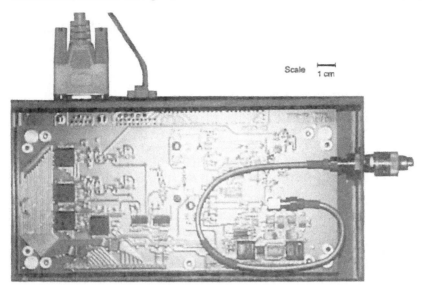

Scale |——| 1 cm

Figure 10-8. Miniaturized time domain reflectometer (Dowding et al., 1996).

MEMORY

A major advantage of TDR technology is the remote monitoring capability. This requires a user to address two of the most basic issues when dealing with data acquisition: 1) the choice of an appropriate spatial resolution, and 2) data acquisition interval. Digital TDR waveforms are acquired as a series of data blocks and the resolution selected will control the number of blocks required to interrogate an entire transducer cable and the range of amplitude in each block. Typically, a TDR waveform is acquired incrementally in data blocks consisting of 251 data points with full-range amplitudes ranging from 10 to 1000 mρ. Measurement resolution must be selected to provide acceptable distance between data points as well as making it possible to detect both small and large amplitude reflections. Too low a resolution will result in missing small, narrow TDR reflections, while too high a resolution consumes excess storage space and requires a long time to transmit data via modem. Assume that transducer cables are being interrogated and

there are 251 waveform data points per block.   If the desired
distance between data points is 0.04 m, then each data block will
cover

$$(251-1) \; data \; points/block \; * \; 0.04m \; = \; 10 \; m/block. \qquad (10\text{-}1)$$

A typical 100 m (328 ft) long cable installed to measure rock mass
deformation with resolution of 0.04 m requires 10 blocks of data
(100 m/(10 m/block), which is equivalent to

$$251 \; data \; points/block \; * \; 10 \; blocks/cable$$

$$= 2,510 \; data \; points/cable. \qquad (10\text{-}2)$$

An extreme case of a 610 m (2,000 ft) long cable with the same
data density would have

$$(251 \; data \; points/block) \; * \; (610 \; m/cable) \; / \; (10 \; m/block)$$

$$=15,311 \; data \; points/cable. \qquad (10\text{-}3)$$

   Data storage requirements at a remote site are an important
consideration. Each data point is typically stored as a 13-bit word
that requires two 8-bit words or bytes for storage. Thus, a 100 m
long cable requires 5 Kbytes of storage (2 bytes/number x 2,510
data numbers) and a 610 m long cable requires 30 Kbytes of
storage. A typical geotechnical instrumentation site might require
4 cables, which are read twice each day but transmitted once per
week. If the cables are 100 m long and data density is 0.04 m, then
each week

$$2510 \; data \; points/cable$$
$$*4cables *7days *2times/day *2bytes/data \; point$$

$$= 281 \; Kbytes \qquad (10\text{-}4)$$

must be stored. Storage modules that can easily store this amount of data are available with a variety of options to greatly increase the storage capacity. However, the time required to transmit data can be prohibitive, as discussed later in connection with telemetry.

VIEWING TDR WAVEFORMS

TDR measurement of point properties (e.g., soil moisture) requires only pulse travel time along the probe. This travel time, or possibly moisture content which has been computed with a programmed algorithm, can be communicated as a single number. This single number display is the objective of many packaged systems in which a user simply connects a probe and then pushes a button to make a measurement.

Very often it is critical to view raw TDR waveforms especially during system installation and during periods of system debugging. This may simply involve viewing the waveform on a cable tester display, using a strip chart recorder with the cable tester, or printing hardcopy of waveforms using data analysis software (see Chapter 11).

TDR monitoring of rock, soil, or structural deformation inherently involves incrementally interrogating long lengths of the transducer cable and then concatenating these files to view a raw waveform to locate and analyze all reflections along the entire cable. With the current state of the art, it is not yet possible to batch process raw waveforms to detect and analyze reflections caused by cable deformation. Discussion of software in Chapter 11 describes steps involved in data analysis for deformation monitoring.

PACKAGED SYSTEMS FOR SOIL MOISTURE

Packaged systems for soil moisture monitoring are available in which the TDR pulser, sampler, and processor are sealed in a small, portable, thermoplastic box. It is designed to be waterproof when the lid is closed, and splashproof when the lid is open (i.e., can be used in the rain). The system available from ESI Environmental Sensors, Inc. interrogates the probe and reduces the segmented probe waveforms (Figure 10-3d) to a numerical data set which can be displayed or can be stored in an internal data logger. It takes approximately 15 seconds per segment for the instrument to interrogate the probe, analyze waveforms, and store moisture measurements. A standard five-segment probe will take approximately 75 seconds to completely measure the moisture for all segments.   The TRASE system available from Soilmoisture Equipment Corp. has boards which can be switched to allow for increased data storage as well as maintaining an internal database of stored TDR waveforms.

COMPUTER CONTROL OR SERIAL COMMUNICATION

The TDR pulser/sampler requires definition of a number of sampling parameters in order to operate. Among these are the location to begin sampling, distance between data points, propagation velocity, and range of amplitude. Furthermore, the pulser/sampler will not store the waveform and a means of storage must be provided. The packaged systems mentioned above are equipped with memory chips and internal dataloggers for this purpose. More commonly, this capability is provided via serial communication with an external computer or datalogger.

The computer or datalogger is programmed to activate the TDR unit, initiate pulse generation, and store the resultant waveform.   Such an arrangement allows great flexibility, including incrementally interrogating the entire length of a transducer cable. For example, 1 m long segments of a 100 m long cable can be incrementally interrogated and the 251 data points per meter per window can be stored along with the acquisition settings.

The remote computer or datalogger is controlled by a program which is downloaded from a base computer connected remotely via telemetry or directly via a serial communications cable. With a program active in the datalogger's memory, it executes the following sequence: turns the TDR pulser power on, acquires TDR waveforms in the first channel, switches to the second multiplexer

Figure 10-9. Multiplexed system used for remote monitoring of strata movement over an abandoned mine.

channel, acquires TDR waveforms in this channel, then repeats the sequence until all channels are interrogated and, finally, turns the pulser power off. This process is totally automatic and requires no operator in the field.

TELEMETRY/CELLULAR DATA ACQUISITION

One of the main advantages of remote datalogger-controlled systems is their telecommunication capability. The installations shown in Figures 10-6 and 10-9 allow data and programs to be downloaded via modem hardwired to existing phone lines. The system shown in Figure 10-9 involved installation of the following equipment in weather-proof enclosures: 1) Tektronix 1502B cable tester with serial communications module, 2) CSI CR10 datalogger, 3) 1200 baud telephone modem, and 4) a 12-volt deep cycle battery trickle charged by a 12-watt solar panel.

Figure 10-10.   System used with radio telemetry for monitoring rock deformation over an active mine.

The installation shown in Figure 10-10 utilized radio telemetry. A critical concern when using telemetry is the time required to transmit stored data. This transmission time can be approximated. First determine the size of the file in bytes. Each byte represents a character and requires 10 bits (8 bits plus 1 start bit and 1 stop bit) for transmission. Add approximately 30% more bits to account for framing and other control characters. Each data block consists of (251 data points + 12 acquisition parameters =) 263 data numbers stored as 2-byte integers. Transmission with a 1200 baud modem would require approximately

263 data numbers/data block *2 bytes/data number

$$= 528 \text{ bytes/data block,} \qquad (10\text{-}5a)$$

and

$$(528 \ bytes/data \ block *10 \ bits/byte *1.3)/(1200 \ bits/sec)$$

$$= 6 \ sec/data \ block. \qquad (10\text{-}5b)$$

For the example of rock deformation measurement mentioned earlier, 281120 bytes (Equation 10-4) would be transferred weekly. Transmission for this week's worth of data requires about

$$(281120 \ bytes/week *10 \ bits/byte *1.3)/(1200 \ bits/sec)$$

$$= 3045 \ sec/week = 51 \ min/week. \qquad (10\text{-}6)$$

The TDR system shown in Figure 10-10 was used to monitor strata movements over an active coal mine. A cable was grouted into a hole 600 m deep and monitored daily. Radio telemetry had to be used to transmit data to a base station. The system was powered by a deep-cycle battery which was trickle-charged by a 20 watt solar panel. Included in the enclosure was an RF modem. A radio was connected to the YAGI antenna mounted at the top of the tower. At the base station, a radio was connected to a YAGI antenna. The radio was connected to an RF modem which was connected to a 1200 baud telephone modem hard-wired to a phone line. The base station was powered by a sealed NiCad battery which was trickle charged by an 8 watt solar panel.

Due to the large data file sizes involved and transmission time required, there were several problems with data retrieval. Extended periods of cold temperatures placed extreme demands on the battery. In addition, the time required to acquire a set of data for such a long cable and the time required to transmit the data caused demands which were too great for the available power. As discussed later, serious consideration must be given to remote power requirements, especially for sites which are distant and where monitoring must be maintained throughout the winter months.

MODEM/PHONE LINE DATA ACQUISITION

The serial communication capability available with a variety of TDR pulser/samplers and dataloggers allows direct digital data acquisition with any personal computer via asynchronous communication. Furthermore, it is possible to remotely control a cable tester with a modem via public telephone lines.

Dowding and Huang (1994a) describe a system which was installed to monitor rock deformation over an active coal mine. The Tektronix 1502B cable tester had a SP232 serial communications module connected to a 2400 baud telephone modem. Both the cable tester and modem were connected to a 120 VAC outlet and powered continuously. Problems developed with 1) power surges during electrical storms, and 2) electrical noise introduced as the high voltage mining equipment advanced past the hole location. The noise problem was primarily due to inadequate grounding of the cable tester and transducer cable.

BATTERY POWER FOR REMOTE MONITORING

Deep-cycle marine batteries are specifically designed to withstand the rigors of marine and recreational vehicle applications, making them the best choice for millions of boating, fishing, RV, and instrumentation applications (Eveready, 1995). Grid alloys and active materials have been designed to withstand the stresses of repetitive cycling without loss of active material. Dense active material is applied to a special deep-cycle grid for maximum ampere-hour capacity.

When a battery falls below a 75% state of charge (Table 10.3), it should be recharged as soon as possible. Also, measures must be taken to prevent overcharging, which results in grid corrosion and water loss, thereby reducing battery life and increasing maintenance (water addition). Batteries in remote locations are typically trickle-charged using solar panels with solid state electronics that prevent overcharging.

To understand the challenge posed by solar panel recharging, such as used with the system shown in Figure 10-10, consider the approximate times necessary to recharge deep-cycle batteries as listed in Table 10.4. If the battery is discharged by 75%, it will require 18.1 hours for recharging with a constant current of 5 amps. Note in Table 10.5 that an 18 W solar panel can provide only 1.5 amps during periods of optimum sunlight. In order to recharge the battery adequately between periods of data acquisition and downloading, it would be better to have a 72 W solar panel. Obviously, periods of low temperature will reduce available battery power, and frequent battery use will overtax the capability of the solar panels, especially during extended periods of cloud cover. The following case history will illustrate measures that can be taken to account for these conditions.

Case History—Dam Monitoring

An illustration of modifications required to power a remote monitoring system is given by Myers and Marilley (1997). Five of eight remote data acquisition units at a dam were powered by 12 V batteries configured to provide 10 amp-hr and recharged by an 18 watt solar panel. The original power systems were provided with the data acquisition units, and the batteries and solar panels were sized for the site conditions by the manufacturer. The battery power systems for all five units performed poorly during the rainy months of late winter and spring. The batteries did not have the reserve capacity to provide power during periods of limited sunlight and the units shut down operation when battery voltage dropped below the minimum operating level until the sun recharged the batteries. In addition, the 18 W solar panels were not large enough to replace power consumed by the units during periods of cloudy days.

Myers and Marilley decided to upgrade the solar-recharged battery power supplies. The downstream gaging station received a 110 amp-hr gel-cell battery and two 50 watt solar panels. The upgrade for the remaining units consisted of a 110 amp-hr gel cell battery and a 77 W solar panel for two units and a 110 amp-hr battery and four 18 W solar panels connected in parallel (equivalent to 72 W of recharge power). The 18 W panels were used to maximize existing hardware. The upgraded power supplies were

Table 10.3. Available battery voltage (Eveready, 1995).

State of Charge	Specific Gravity	Voltage (12 V battery)
100%	1.265	12.7
75%	1.225	12.4
50%	1.190	12.2
25%	1.155	12.1

Table 10.4. Hours required for recharging 12 volt
deep-cycle batteries (Eveready, 1995).

Percent Discharged	State of Charge	Hours to Recharge[1]		
		@5 amps	@10 amps	@15 amps
25%	75%	6.0	3.0	2.0
50%	50%	12.1	6.0	4.0
75%	25%	18.1	9.1	6.0
100%	---	24.1	12.1	8.1

[1] The above recharge times assume 5% more than theoretical actual (i.e., 105%).
This will ensure total recharge of battery plus thorough mixing of electrolyte.

Table 10.5. Current available for trickle charging
a 12 volt system with solar panels.

Solar panel capacity (W)	Current available (amp)
18	1.5
36	3.0
72	6.0

designed in a way to provide roughly 45 days of reserve battery
capacity and a factor of safety of 3.0 for the solar recharge. All of
the upgraded power supplies performed well during the subsequent
winter and spring seasons.

DETAILED EXAMPLES

An important parameter for any type of instrumentation system is the cost involved. There are fixed costs for hardware and variable manpower costs for installation, data acquisition, and data interpretation. For purposes of illustration, detailed costs of typical systems used for monitoring rock deformation and soil moisture are given in Tables 10.6 and 10.7. However, only the fixed hardware costs are listed.

Rock Deformation

The system shown in Figure 10-9 involved lead cables connected to three coaxial transducer cables and brought to a central location where a utility pole was installed. The data acquisition system was installed on this pole within enclosures. The lead wires were connected to the multiplexer. The multiplexer is connected to a TDR cable tester and a datalogger which is connected to a storage module and modem. The datalogger was programmed to turn on the cable tester, interrogate each cable, store data in the storage module, and then turn off the cable tester. Data was downloaded from the storage module via phone line. The costs for this system when installed in 1995 are itemized in Table 10.6.

Soil Moisture

Multiplexed, remote monitoring of soil moisture is essentially the same as the procedure used for rock deformation except that moisture probes (e.g., Figure 3-1) are used instead of transducer cable. In the case of soil moisture monitoring, waveforms are processed using algorithms programmed into the data logger so that it is necessary to store only a few values for each probe. Consequently, a user can accumulate many weeks of data before downloading without having to consider memory storage limitations. The system shown in Figure 10-6 was installed at a site where AC power was made available. The costs for this system when installed in 1995 are itemized in Table 10.7.

Table 10.6. Detailed Example—Rock Deformation.

	Transducers/Probes		
coaxial monitoring cable	CommScope P3-75-875 CA	2000ft @$0.70/ft	$1,400
connectors	Bilbert Eng. GRS-875-BAFF	3 @ $27.25 ea	$82
low-loss lead cable	CommScope F1160BEF	1000 ft reel	$158
connectors	Gilbert Eng. GF11AHS460	3 @ $2.50 ea	$8
		SUBTOTAL	$1,648
	TDR Pulser/Sampler		
cable tester	Tektronix 1502B	1 ea	$9,500
communication interface	CSI SDM1502	1 ea	$300
power control	CSI PS1502B	1 ea	$130
		SUBTOTAL	$9,930
	Remote Control and Data Acquisition		
grounding kit	CSI UTGND	1 ea	$100
enclosure for cable tester	ENC TDR plus supressor	1 ea	$760
datalogger	CSI CR10	1 ea	$1,090
keypad for datalogger	CSI CR10KD	1 ea	$275
PROM for datalogger		1 ea	$300
RS232 interface for datalogger	CSI SC32A	1 ea	$145
storage module	CSI SM716	1 ea	$590
RS232 interface	CSI SC532	1 ea	$180
coaxial multiplexer w/enclosure	CSI SDMX50	1 ea	$525
connector cable for multiplexer	SDMCBL-L and COAXTDR-L	3 m (10 ft)	$50
telephone modem	CSI DC112	1 ea	$295
20 watt solar panel	CSI MSX20R	1 ea	$470
deep cycle battery and enclosure		1 ea	$60
software	CSI PC208	1 ea	$200
		SUBTOTAL	$5,040
	TOTAL FOR SYSTEM		$16,618

Table 10.7. Detailed Example—Soil Moisture Monitoring.

Transducers/Probes			
soil moisture probes	CSI CS600	8 @ $85.00 ea	$680
TDR Pulser/Sampler			
cable tester	Tektronix 1502B	1 ea	$9,500
communication interface	CSI SDM1502	1 ea	$300
power control	CSI PS1502B	1 ea	$130
		SUBTOTAL	$9,930
Remote Control and Data Acquisition			
grounding kit	CSI UTGND	1 ea	$100
enclosure for cable tester	ENC TDR plus supressor	1 ea	$760
datalogger	CSI CR10	1 ea	$1,090
keypad for datalogger	CSI CR10KD	1 ea	$275
RS232 interface for datalogger	CSI SC32A	1 ea	$145
storage module	CSI SM716	1 ea	$590
RS232 interface for storage module	CSI SC532	1 ea	$180
coaxial multiplexer w/enclosure	CSI SDMX50	1 ea	$525
connector cable for multiplexer	SDMCBL-L and COAXTDR-L	3 m (10 ft)	$50
telephone modem	CSI DC112	1 ea	$295
software	CSI PC208	1 ea	$200
		SUBTOTAL	$4,210
	TOTAL FOR SYSTEM		$14,820

PENDING DEVELOPMENTS

Tables 10.6 and 10.7 are segmented into the following elements: 1) transducers/probes, 2) TDR pulser/sampler, and 3) control and data acquisition.   This allows the reader to readily identify which elements involve the greatest costs. As stated in the introduction of this chapter, the state of the art in instrumentation is dynamic and is expected to be constantly changing. The components and costs in these tables should be considered as a snapshot. TDR pulser/ samplers are currently being beta-tested which will be less expensive.   Packaged systems will ultimately be produced that will decrease the costs for not only the pulser/sampler but also system control and data acquisition.

Chapter 11
SOFTWARE

There is a wide variety of software available for TDR data acquisition and interpretation. One category of software is designed specifically for control of pulser/samplers, a second category is designed specifically for analysis of TDR waveforms, while a third category serves both functions. The summary in Table 11.1 is not extensive but provides an overview of many programs that are available. The summary in Table 11.2 indicates whether a program can be used for data acquisition and/or analysis in the context of geomeasurements.

In order to appreciate the capabilities of available software, common details of programs will be described. Discussion begins with general control and acquisition software, followed by software that has been specifically developed for analysis of soil moisture and deformation measurements.

GENERAL CONTROL/ACQUISITION SOFTWARE

SP232 - Laptop Computer Control

The Tektronix SP232 Host Application Program is supplied with the SP232 serial module (Tektronix, 1989b). It controls the digital 1502 TDR cable tester through a PC (usually a field-portable laptop) connected to the module either directly or via modem as shown in Figure 10-1. Manual control (i.e., knobs on the front panel) of the TDR is disabled and waveforms are acquired with software settings. The grid shown in Figure 11-1 appears simultaneously on the TDR screen and computer monitor. Waveforms are acquired by modifying settings as desired and downloading them to the TDR, which then sends digital waveform data back to the computer.

Table 11.1. TDR software summary.

Name	Supplier	WWW Address	Operating System	System Requirements	Controls	Cost
NUTSA	GeoTDR, Inc	http://www.geotdr.com	DOS	386 PC, 0.6 MB		$50
TRAP						$500
WinTDR	Utah State University	http://tal.agsci.usu.edu/home/ron/web/tdr/win_tdr.html	Win 3.x	386 PC, 5 MB	Tektronix 1502	free
ViewPoint	ESI Environmental Sensors, Inc.	http://www.esica.com	DOS	486 PC, 0.6 MB	MP-917	w/Moisture Point
SP232 Host Program	Tektronix, Inc.	http://www.tek.com	DOS	386 PC, 0.2 MB	Tektronix 1502	w/SP232 module
PYETDR	CSIRO	zegelin@python.enmech.csiro.au	DOS	386 PC, 0.5 MB	PYETDR	
PC208	Campbell Scientific, Inc	http://www.campbellsci.com	DOS, Win95	486 PC, 1 MB	CR10	$220
WINTrase	Soilmoisture Equipment Co.	http://www.soilmoisture.com	Win95	486 PC, 1 MB	TRASE	$1500
SUPER STORE	Riser-Bond	http://www.riserbond.com			Model 1270 TDR	
WAVE-VIEW			Win95			
700WS-TDRWin	Signatel	peter@lociscio.com	Win95		700 series TDR	

Table 11.1. (continued)

Name	Supplier	WWW Address	Operating System	System Requirements	Controls	Cost
Smart Rain	Smart Rain Corp. Inc.	http://www.SmartRain.com	Win95	586 32 MB RAM	Smart Rain sensors	w/Smart Rain
TACQ	Dynamax	http://www.dynamax.com	DOS	386 PC, 0.6 MB	Tektronix 1502	
LOM/RS	Easy Test	henryk@gaja.ipan.lublin.pl	DOS	386 PC, 0.5 MB	LOM	w/LOM
TDRPlot	KANEGeoTech	http://www.kanegeotech.com	Win 3x			w/short course
PCI-3100	CM Technologies	ecadusa@aol.com	Win 3.x, DOS		PCI-3100	
SMCAL	Imko	http://www.alive.de/imko/	DOS	386 PC, 0.5 MB	TRIME	
PALCOM	PermAlert		Win3.x, DOS	PC	PAL-AT	
REFLEX 2.0	CANMET-MRL	taston@nrcan.gc.ca	Win 3.x	386 PC, 2 MB		
E 2790	Edgecumbe Instruments	support@metrohm.demon.co.uk	Win95	PC	Metrohm E27	

Table 11.2. TDR software applications.

Name	Supplier	Acquisition				Analysis			
		soil moisture	conduc- tivity	water level	defor- mation	soil moisture	conduc- tivity	water level	defor- mation
NUTSA	GeoTDR, Inc						x	x	x
TRAP	GeoTDR, Inc					x	x	x	x
WIN_TDR	Utah State Univ.	x	x			x	x		
ViewPoint	ESI Environmental Sensors	x				x			
SP232 Host Application Program	Tektronix	x	x	x	x	x	x	x	x
PYETDR	CSIRO	x	x			x	x		
PC208	Campbell Scientific, Inc	x	x	x	x	x	x		
WINTrase	Soilmoisture Equipment Co	x	x			x	x		
700WS-TDRWin	Signatel	x	x	x	x	x	x	x	x
SUPER STORE	Riser-Bond	x	x	x	x				
WAVE-VIEW						x	x	x	x
Smart Rain	Smart Rain Corp. Inc.	x	x			x	x		

Table 11.2. (continued)

Name	Supplier	Acquisition				Analysis			
		soil moisture	conduc- tivity	water level	defor- mation	soil moisture	conduc- tivity	water level	defor- mation
Vadose TDR	Dynamax	x	x			x	x		
LOM/RS	Easy Test	x	x			x	x		
TDRPlot	KANE GeoTech					x	x	x	x
PCI-3100	CM Technologies	x	x	x	x	x	x	x	x
INMEWA	Imko	x	x			x	x		
PALCOM	PermAlert		x	x			x	x	
REFLEX 2.0	CANMET-MRL					x	x	x	x
E 2790	Edgecumbe Instr.	x	x	x	x	x	x	x	x

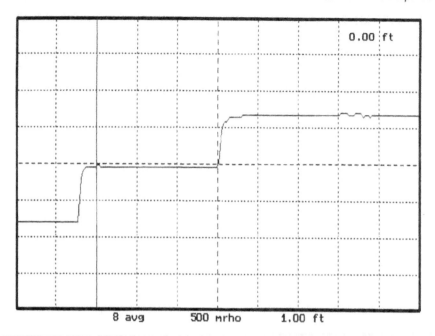

```
Get, Set, Read, Write, Info, Alt, Diff, Erase, Quit, More?
```

Figure 11-1. SP232 Host Application Program (from Tektronix, 1989).

Acquisition Settings

The acquisition settings file shown in Table 11.3 is an ASCII file
created with the Tektronix SP232 program or, more conveniently,
using any DOS editor. The settings include

    1) number of waveforms to be averaged,
    2) vertical sensitivity,
    3) horizontal sensitivity,
    4) propagation velocity,
    5) distance at which interrogation should begin,
    6) vertical position of waveform on screen, and
    7) location of cursor on screen.

The settings file may be saved and modified for several different
applications and projects. Most important, for purposes of deform-

Table 11.3. SP232 settings file.

First screen setting   (acquire data from -2 ft to 8 ft)

instr_id: 1502   <----- instrument ID
averages: 8   <----- number of wave forms to be averaged
vertical: 6078   <----- vertical offset
gain: 20.0000 mrho   <----- vertical scale (mrho/vertical div)
ddiv: 1.0000 feet   <----- horizontal scale (distance per division)
vp: 0.66   <----- propagation velocity along cable
cpos: 0   <----- cursor position (cursor location on screen)
cdist: -2.0000   <----- cursor distance (position of 1st data point)
maxhold: off
pulse: off
singsweep: off
dspohms: off
buttons: 1

Second screen setting   (acquire data from 8 ft to 18 ft)

instr_id: 1502
averages: 8
vertical: 6078
gain: 20.0000 mrho
ddiv: 1.0000 feet
vp: 0.66
cpos: 0
cdist: 8.0000   <----- set cursor at 8 ft
maxhold: off
pulse: off
singsweep: off
dspohms: off
buttons: 1

ation monitoring, the settings file can be thought of as a script which allows incremental interrogation of transducer cables as discussed below.

Remote Use

Once a settings file is created, it can be employed to remotely acquire waveforms by establishing modem communication with a remote 1502 TDR using communication software (e.g., PRO-COMM PLUS, PCAnywhere, etc.). Once communication is established between the remote TDR and base computer, the SP232 program performs the following:

1) acquires the first screen waveform (i.e., 251 data points) at settings specified in the file, displays the TDR waveform, and downloads all information to a specified output file,

2) acquires the second screen waveform, displays the second TDR waveform, and downloads all information of the second screen to output file (i.e., appends it to the specified output file), and

3) continues to acquire third waveform, fourth waveform, etc. until all the specified waveforms are acquired and downloaded to the output file. At present this software can be used only at a baud rate of 1200.

An example SP232 settings file for two screens of data is given in Table 11.3. Incremental waveforms were obtained from a cable 9 m (30 ft) long using this settings file. The block of data obtained for the first screen only is shown in Table 11.4. The data block contains three parts: a) title, b) settings, and c) waveform data. The waveform data consists of 251 values of the TDR reflection coefficient. The distance interval between data points is 30 ft /251 = 0.125 ft. The first data point is at -2 ft, which is the specified cursor distance (cdist) in the settings file, and the last point is at (-2 ft + [2 ft/screen div][10 screen div] =) 8 ft.

Use of SP232 allows capture of raw waveform data for any TDR measurement: soil moisture, deformation, water level, etc. Thus, if the investigator is willing to analyze TDR waveform signatures by graphing the digital data, no other programs are needed. Specialized software discussed later in this chapter allows more flexibility when analyzing and interpreting the raw data.

Table 11.4. Output file for first screen.

Title	date: Thu May 10 13:37:24 1990 * notes: *
Settings	instr_id: 1502 averages: 8 vertical: 6078 gain: 20.0000 mrho ddiv: 1.0000 feet vp: 0.66 cpos: 0 cdist: -2.0000 maxhold: off pulse: off singsweep: off dspohms: off buttons: 1
Waveform data	waveform: acquired seq: 1 0 0 0 251 5

```
              0   0   0   0   0   0   0    0   0   0   0   0   0   0   0   0
              0   0   0   0   0   0   0    0   0   0   0   0   0   0   0   0
              0   0   0   0   0   0  32   55  52  58  58  60  63  62  66  81
             78  73  69  64  62  61  61   62  60  59  58  58  58  59  58  59
             59  58  58  58  59  58  58   58  59  59  59  59  59  59  59  59
             59  59  59  59  59  60  60   60  60  59  60  60  60  60  60  59
             59  59  60  60  60  60  60   60  60  60  59  59  60  60  61  61
             60  60  60  61  61  60  60   60  60  61  62  69  75  65  21   4
             28  46  61  65  65  66  65   64  64  65  63  64  67  68  68  58
             45  49  56  61  64  63  63   65  66  67  67  66  56  43  47  56
             61  64  59  48  51  57  61   64  64  63  64  65  65  65  67  67
             66  66  66  66  67  67  66   66  66  66  66  67  67  65  55  49
             54  61  64  64  64  64  65   66  67  67  66  67  67  66  64  64
             67  67  67  66  66  67  67   67  64  63  68  68  66  67  68  67
             66  65  66  67  67  66  62   54  53  58  60  64  65  64  64  65
             67  68  67  66  67  67  66   64  66  69
```

CSI PC208 - Datalogger Control and Multiplexing

Just as a computer can be connected to a TDR pulser/sampler, so
can a programmable data logger, which allows remote, continuous
data acquisition. A program can be stored and executed by a
datalogger to automatically perform the functions of logical
control, measurement, data processing, and data storage. In
general, a typical programming sequence requires that 1) the
execution interval be specified, 2) sensor output be measured, 3)
measurements be processed, and 4) results be stored. This
sequence can be employed for implementing all the applications of
TDR using a single datalogger and TDR pulser/sampler connected
to many transducer cables and probes through a multiplexer. For
example, a datalogger program can be written for three channels of
water level monitoring, two channels of deformation monitoring,
and three channels of soil moisture monitoring. The variety of
instructions available allows selection of measurement, processing,
data storage, and control sequences that precisely fit specific project
needs.

Acquisition Parameters

The most common data logger for TDR measurements is the CR10
(Campbell Scientific, Inc., 1992c), which is programmed with
PC208 software (Campbell Scientific, Inc., 1992b). A typical data
block structure is shown in Table 11.5. A total of 263 raw data
numbers are defined per block of output, which is similar to the
Tektronix data block shown in Table 11.4. The CSI data block
structure starts with a program statement number (108 in this case),
followed by 11 data values (ending in 53264). Julian year, day,
hour-minutes, and seconds are times at which the measurement is
made. Battery voltage and module temperature are parameters that
monitor the datalogger's voltage and enclosure temperature. The
data window number indicates the position of the acquired data
block within a series of blocks acquired when incrementally
interrogating a transducer cable. Cursor distance, distance between
points, and gain are acquisition settings that control data resolution
or density. Offset is the value of the vertical position of the wave-

Table 11.5. Sample output file acquired using CSI PC208 software.

Parameters and Data

108,1993,76,1450,28.38,13.2,22.7,1,37.5,.01,50,53264,

4064,4074,4122,4017,4112,4078,4115,4060,4189,4113,4130,4132,4094,
4094,4075,4116,4068,4101,4039,4062,4110,4029,4077,4148,4099,4181,
4169,4183,4119,4218,4158,4193,4165,4179,4151,4160,4173,4110,4174,
4182,4179,4165,4157,4163,4161,4171,4192,4165,4167,4145,4244,4135,
4139,4149,4158,4161,4157,4219,4113,4200,4223,4237,4175,4237,4196,
4237,4219,4209,4178,4154,4164,4161,4122,4068,4133,4099,4100,4129,
4189,4124,4176,4175,4174,4080,4153,4122,4128,4094,4107,4114,4140,
4125,4139,4158,4166,4188,4178,4179,4206,4210,4146,4148,4122,4085,
4140,4162,4074,4161,4145,4178,4151,4286,4167,4210,4243,4218,4238,
4244,4254,4213,4240,4177,4191,4166,4125,4174,4160,4168,4180,4211,
4235,4192,4275,4222,4255,4242,4212,4240,4228,4282,4205,4238,4168,
4184,4183,4191,4167,4220,4223,4202,4254,4252,4197,4298,4251,4243,
4212,4179,4152,4183,4189,4082,4167,4140,4180,4173,4260,4200,4209,
4185,4198,4209,4287,4272,4270,4228,4233,4237,4206,4219,4233,4182,
4192,4208,4251,4125,4216,4203,4195,4273,4249,4281,4206,4278,4171,
4225,4183,4154,4181,4105,4169,4189,4169,4180,4148,4229,4195,4276,
4191,4189,4251,4214,4273,4176,4264,4178,4186,4153,4154,4168,4150,
4170,4159,4206,4222,4204,4233,4277,4260,4258,4255,4269,4263,4284,
4254,4278,4196,4223,4226,4278,4179,4283,4274,4294,4286,4276,4252,
4246,4329,4267,4309

Value	Parameter Description
108	number of program table and statement that caused execution of data storage
1993	Julian year
76	Julian day
1450	hour-minutes
28.38	seconds
13.2	battery voltage
22.7	module temperature ($^{\circ}$C)
1	window number
37.5	cursor distance (m)
.01	distance between points (m)
50	gain (mρ)
53264	offset

Following the above 12 data values are the 251 waveform data values.

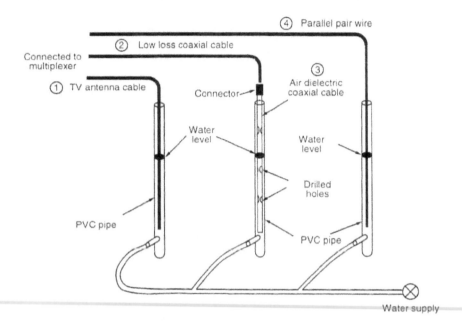

Figure 11-2. Setup used for demonstration of monitoring multiple transducer cables.

form within the window which is similar to "vertical" in Table 11.3.

Example Interrogation of Multiple Transducers Cables

A laboratory water level measurement test was conducted to demonstrate the multiplexing capability and serve as a basis for an example datalogger program. A schematic diagram of the demonstration is shown in Figure 11-2. The test employed three different types of cables/wires. The first was a 30 m (100 ft) long TV antenna cable, the second was a 80 m (250 ft) long coaxial lead cable with a 2.5 m (7.5 ft) long air-dielectric coaxial cable as a transducer, and the third was a 30 m (100 ft) long parallel pair antenna wire. Holes were drilled in the air-dielectric coaxial cable every foot to ensure water could enter and drain rapidly. The cables/wires were connected to multiplexer channels 1, 5, and 8, respectively. The multiplexer was connected to a Tektronix 1502B TDR and CR10 datalogger with an external storage module. See Chapter 9 for a discussion of the study results obtained using the setup shown in Figure 11-2.

During the test, the water level was raised to the top of the pipes, then lowered at a slow, constant rate. The datalogger program was set to acquire waveforms every 10 minutes and data was downloaded from the storage module once an hour. The program performed the following sequence:

1) execute the main program continuously,
2) at a specified time interval, turn on the cable tester,
3) establish serial communication between the datalogger and cable tester,
4) begin the measurement loop for multiplexer channel 1,
5) store the datalogger battery voltage in datalogger memory,
6) store the datalogger temperature in datalogger memory,
7) acquire one screen of a TDR waveform,
8) store the settings information and 251 waveform data point values in datalogger memory,
9) output all data from datalogger memory to the storage module,
10) increment the data block number and repeat steps 5 to 9 until the specified number of blocks has been acquired, then
11) repeat steps 4 to 10 for channel 5 and channel 8, and finally
12) turn off the cable tester to conserve power and wait for the next specified time for measurement.

The acquisition time for one channel in this example was estimated to be less than 2 minutes. This time is dependent on the number of data blocks to be acquired. This, in turn, is dependent on the transducer cable length and desired distance between data points. Refer to the discussion of storage memory in Chapter 10.

SPECIALIZED SOFTWARE

Soil Moisture —Acquire and Interpret Waveforms

Measurement of soil moisture and conductivity can be accomplished
with real time interpretation of TDR waveforms. Rather than
storing an entire waveform, algorithms have been developed which
compute the travel time along a probe as well as the magnitude of
the reflection from the end of the probe. The computed values are
employed to compute moisture content and conductivity, which
may be the only values stored. If desired, complete waveforms
may be acquired and stored using the Tektronix SP232 software or
CSI PC208 software, but other programs will be discussed to
illustrate other features that are available.

Parallel Rod Probes: WinTDR and PYELAB

WinTDR and PYELAB are examples of available interpretive
programs. WinTDR was written by the Soil Physics Group at Utah
State University (Hubscher and Or, 1996). This program was
developed to measure volumetric water content using the Tektronix
1502B/C series TDR. Since its conception in 1993, it has been
modified to analyze electrical conductivity, employ a variety of
commercial multiplexers, and give the user an enhanced ability to
interact with the system. The intent of the program has been to
create a friendly Windows interface with accurate and efficient
analysis under a variety of conditions. PYELAB is part of the
system developed by CSIRO to allow user-friendly determination
of soil moisture and conductivity. Both programs determine the
bulk dielectric constant from the apparent length, l_a, shown in
Figure 11-3. Since the actual length of the probes is fixed, the
apparent length can be converted to dielectric constant from which
volumetric water content is computed by applying Topp's equation
(Equation 3-7).

Current features of Win TDR include 1) use the computer's
internal clock to make timed readings, 2) use of multiplexers, 3)
self-calibration methods to increase accuracy, 4) an open interface

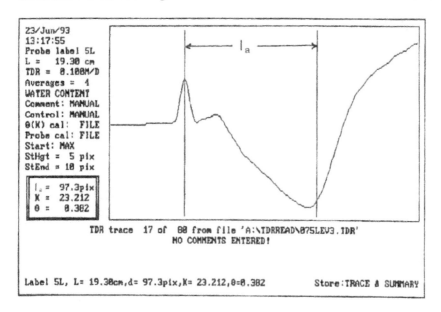

Figure 11-3. Determination of soil moisture with PYELAB software (from CSIRO, 1991).

to make important data visible to the user, and 5) the ability to quickly and easily alter the information being analyzed. The system is also provided with features to change the third-order polynomial coefficients of Topp's equation with other coefficients. The raw waveform data and analysis results can be saved to files.

The user can perform readings either automatically or manually. The program will produce the following: a) Location—the multiplexer number and channel number of the probe being read, b) Date—indicates the current hour:minute:second day/month/ year of the reading, c) Bulk Dielectric—the calculated bulk dielectric constant, d) Analyze Water Content—determines the current water content, and e) Analyze Salinity—determines the bulk conductivity based on attenuation of the waveform similar to PYELAB as shown in Figure 11-4.

The WinTDR users manual can be downloaded from the Utah State University web site (Table 11.1). The software is constantly being modified, but the manual is updated only periodically and the most recent version should be obtained. WinTDR

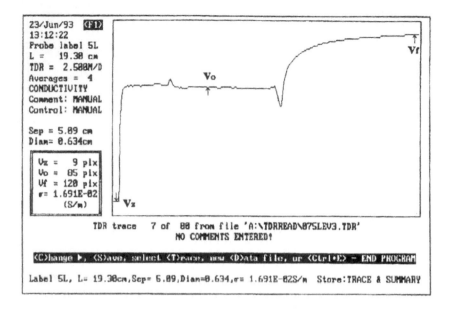

Figure 11-4. Determination of bulk conductivity with PYELAB software (from CSIRO, 1991).

requires the following hardware: a) PC 386 or better running under Windows 3.x or Windows 95, with 5 MB of hard disk space, b) Tektronix 1502B (or compatible) TDR unit, and c) a 3-prong soil moisture probe.

Segmented Probes: ViewPoint

ViewPoint software controls data acquired from segmented MoisturePoint™ probes (ESI Environmental Sensors, Inc.) and thus requires special discussion.   These probes have switching diodes which divide them into segments (Figure 10-3c).   A diode is an electrical device that conducts current in only one direction.   These diodes are employed to controllably introduce short circuits (impedance discontinuities) across a transmission line as shown in Figures 11-5a,b.   By controlling which diodes are on/off, specific segments of the probe can be examined.

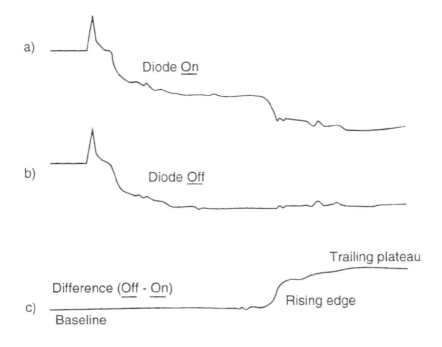

a) Diode On

b) Diode Off

c) Difference (Off - On)

Baseline

Rising edge

Trailing plateau

Figure 11-5. TDR waveforms displayed with ViewPoint demonstrating diode switching. Note that all reflections prior to the diode are eliminated in the difference waveform (from ESI Environmental Sensors, 1995).

Diode switching control

A particular shorting diode can be either off (not conducting) or on (conducting). For each condition, a TDR pulse will return a different reflected pulse, but they will be different only past the point where the diode is located (Figure 11-5c). The reflected pulse with the diode on (transmission line shorted) will exhibit a downward or negative step at the diode location. The reflected pulse with the diode off will not examine this. Common features of both waveforms cancel out, and only the differences between the waveforms are seen. The difference eliminates features that may be present in both of the reflected waveforms, but not caused by the diode short (e.g., an impedance bump caused at the probe connector). A typical difference function will have the following characteristics: a flat, smooth baseline, a sharply rising edge, and a

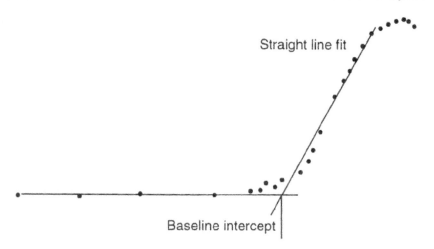

Straight line fit

Baseline intercept

Figure 11-6. Determination of intercept between baseline and straight line fit along rising edge (from ESI Environmental Sensors, 1995).

relatively flat trailing upper plateau.

Difference functions are employed to pinpoint the position of a diode. The ESI MP-917 pulser/sampler uses special circuitry to measure and acquire diode difference functions (and raw TDR waveforms as well). During probe scans, examination of the rising edge of the difference function allows the MP-917 to accurately determine the position of a diode in the time domain. This is done by fitting a straight line to the rising edge of the difference function, and extrapolating the line down until it intersects the baseline as shown in Figure 11-6. The intercept point on the baseline is the diode's position in time. By following this procedure at each end of a probe segment, the propagation time of a pulse between two diodes can be defined. This propagation time, or time delay between diodes, is compared by the MP 917 to calculate the moisture content of the soil around the probe, at the elevation between the two diodes.

In field use, the MP-917 automatically performs all of the above steps, and displays only the resulting moisture content for each segment of the probe. ViewPoint software provides a means to examine in detail the waveforms that are used by the MP-917 to calculate moisture readings. Raw TDR waveforms, diode difference functions, and complete probe scans can be examined.

Types of scans

ViewPoint displays the waveforms that the MP917 uses to measure moisture content. There are two general types of waveforms that can be displayed: diode waveform scans and probe waveform scans. Diode scans show in detail the raw TDR waveforms for a selected diode. A probe scan is the collection of all diode difference functions required to produce a complete moisture profile with a MoisturePoint™ probe. The MP-917 gathers all the data it needs to make a moisture measurement for each segment of a probe, performs analysis of this waveform data, and produces the moisture results with an algorithm presented by Hook and Livingston (1996).

ViewPoint collects all of this information and displays it graphically on the screen. The difference waveforms are plotted, along with straight lines fitted to the rising edges of the waveforms as shown in Figure 11-7. Vertical lines extend down from each point where a slope line intercepts a baseline of the difference functions. These vertical lines serve to mark the diode's position in the time domain, and also act as boundaries between segments. The moisture data that is calculated for each segment is displayed between these lines. The upper number is the raw time difference. The units for this time difference are the internal MP-917 time base units used by the fine delay generator. The middle number is the moisture in %, and the bottom number is the time delay in nanoseconds.

Deformation—Analyze Changes in Raw Waveforms

Deformation monitoring is very different than soil moisture monitoring because, rather than analyzing data at a predetermined location, it is necessary to acquire complete waveform data for long transducer cables. The waveform must be acquired in segments and several blocks of data must then be concatenated to allow analysis of the waveform for the entire cable. Random noise can appear on TDR waveforms and algorithms must be assembled or developed to eliminate this noise. While this software is still

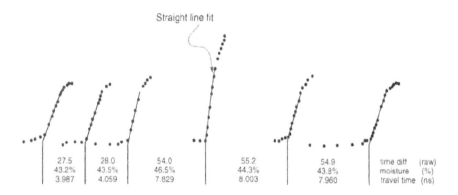

Figure 11-7. Example probe scan. For each segment, the top value is the raw time difference in internal time base units, the middle value is moisture content, and the bottom value is the travel time in nanoseconds (from ESI Environmental Sensors, 1995)

being developed, a first-generation version is available for analysis of changes in raw waveforms.

NUTSA

TDR waveform data may be processed using a wide variety of approaches. The simplest approach is simply to compare strip chart hard copy of waveforms. The next level of processing includes common scaling of waveforms for simple visual comparison on a computer monitor. Finally, analytical programs can be developed to interactively analyze individual reflections or compare one waveform with another. These computerized analytical capabilities have been implemented in NUTSA (Huang et al., 1993). It is designed to allow visual comparison of up to three TDR waveforms quickly and easily as well as converting raw digital waveforms to ASCII files that can be employed with other spreadsheet and plotting programs. This approach maximizes the efficiency of human decision making and visual acuity by providing a rapid means of comparison and analysis.

Development of NUTSA was motivated by the need to analyze and interpret large numbers of TDR records obtained when monitoring several long cables on a daily basis. The location, amplitude, and width of changes in the waveforms can be compared, quantified, and recorded digitally upon visual detection of a significant difference. This approach avoids development of signal interpretation algorithms and facilitates human judgement by providing automatic visual comparison, while still making it possible to quantify differences.

Waveform data formats

When a user invokes the load waveform files option in the main menu as shown in Figure 11-8, a pop-up window appears that prompts the user to input up to three TDR waveform files for analysis. Valid input files can be created with either the Tektronix SP232 or Campbell Scientific, Inc. PC208 software discussed earlier. The Tektronix SP232 file format consists of three parts for each screen: 1) title, 2) settings, and 3) waveform data as shown in Table 11.4. NUTSA reads all this information for the first screen, then only waveform data for subsequent screens. The CSI PC208 file format consists of a series of values including waveform ID, year, hour/min, seconds, etc., followed by waveform data as shown in Table 11.5. NUTSA reads all this information, then computes the cursor distance and distance between points based on the value for propagation velocity input by a user as shown in Figure 11-9. NUTSA then adjusts the vertical offset to account for auto-ranging by the CR10 datalogger as shown in Figure 11-10. Originally, the CSI CR10 datalogger acquired waveforms with a default propagation velocity, $V_p = 0.99$. This value can now be varied and it is necessary to adjust V_p to an appropriate value for the combined lead and transducer cable. The program prompts a user to enter V_p and a pop-up help window is displayed for reference, as shown in Figure 11-9. The value is selected so that the displayed horizontal scale (i.e., distance) will be physically meaningful. This value may be well outside the range of values listed in Figure 11-9.

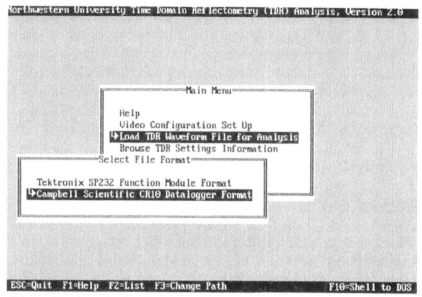

Figure 11-8. Load waveform files for analysis by first selecting the correct data file format.

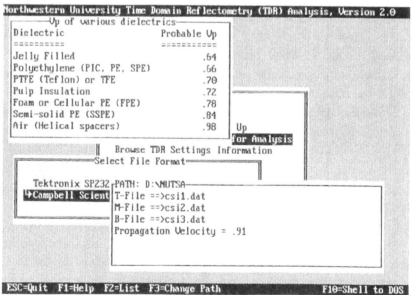

Figure 11-9. Input up to three waveform files for analysis and specify the propagation velocity.

Creation of scratch files and concatenation of data blocks

NUTSA reads all waveform data screen by screen and creates scratch files (i.e., TFILE.TMP, MFILE.TMP, BFILE.TMP) to decrease the time required to redraw waveforms. For example, when working with files acquired with PC208 software, NUTSA performs the following based on true V_p: 1) adjusts the parameter "cursor distance," 2) adjusts the parameter "distance between points," 3) adjusts relative vertical offset, and 4) accounts for the overlapping of each screen by storing 251 data points for the first screen, but ignoring the first data point of each subsequent screen. A complete waveform scratch file is created which reduces the time required to redraw waveforms.

Unlike Tektronix's SP232 algorithm, in which all data blocks have the same vertical offsets, those for data blocks acquired using CSI's algorithm are not the same. The datalogger acquires TDR signals with the optimum offset for each data block as illustrated in Figure 11-10a. NUTSA had to be modified to account for CSI's auto-ranging system. As shown in Figure 11-10b, a series of data blocks can be concatenated such that the last data point of a block overlaps the first data point of the subsequent block. NUTSA adjusts offset values for successive data blocks so that all have a common baseline for the entire interrogated cable. If the acquired waveform signature exceeds the screen limits as shown in Figure 11-10b, NUTSA allows scaling and positioning so the entire waveform is within the window limits as shown in Figure 11-10c.

Identification of waveform changes

NUTSA was developed for visual analysis of changes in TDR waveforms. The functionality of NUTSA options is described in Table 11.6. The menu window along the left side of the screen shown in Figure 11-11 provides choices for changing display parameter values and for invoking "Diff" (difference between waveforms) and "Zoom" (change horizontal scale). Units for each parameter are dictated by the settings chosen for data acquisition. The value of Vscale, Sdist, and Edist, in force when waveforms are initially viewed, is dictated by the T-File settings.

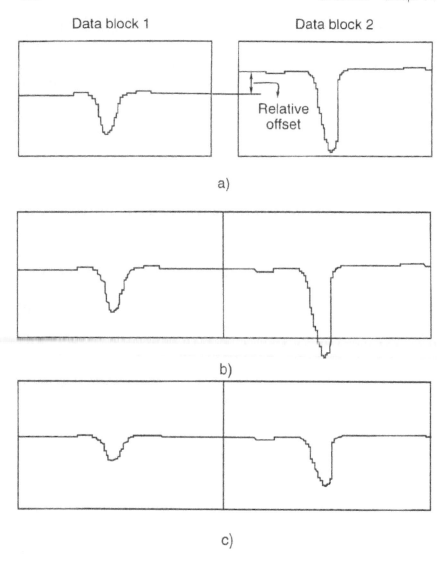

Figure 11-10. Adjustment of vertical offset and concatenation of TDR waveform data blocks; (a) waveform data acquired by data logger with optimal vertical offset; (b) vertical offset adjusted to form continuous waveform; (c) rescale adjusted waveform.

If a cable is long, small spikes or changes cannot be seen when the entire waveform is displayed due to limitations of screen resolution. The alternative is to zoom or magnify an area of interest. NUTSA provides two approaches: (1) Highlight "Sdist"

Table 11.6. Parameters used in NUTSA for analysis of TDR waveforms.

HOT Keys	
<ESC>	escape
<F1>	online help
<F2>	write waveform to an ASCII file
<F7>	redraw waveforms after changing display parameters
<F8>	reset to initial display parameters
<F10>	invoke cursor in waveform display

Waveform Display	
Vscale	vertical scale (mrho/division)
Hscale	horizontal scale (feet, meters, or arbitrary)
Tindex	vertical position of waveform in top window
Mindex	vertical position of waveform in middle window
Bindex	vertical posisition of waveform in bottom window
Sdist	distance value at start of window
Edist	distance value at end of window
Zoom	invoke option to zoom into a portion of waveform (use cursor to change Sdist and Edist)

Waveform Analysis	
Peak Loc	location of cursor
Amplitu	amplitude of waveform at cursor location
Diff	invoke option to calculate the difference between two waveforms
<Ctrl><D>	compute magnitude and width of spike in difference waveform
PWidth	width of spike in difference waveform
<Ctrl><S>	save information about spike in difference waveform
Write	write waveform data to an ASCII file for use use with other spreadsheet and plotting software

and "Edist" and enter the desired values, or (2) in the "Zoom" feature, use a cursor to select desired beginning and ending locations.

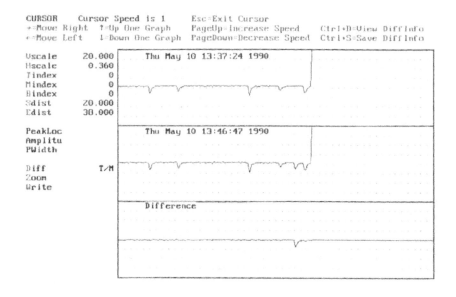

Figure 11-11. The difference between two waveforms can be automatically displayed.

Quantifying waveform changes

The cursor (vertical line shown in Figure 11-12) can be moved with arrow keys to any location in any of the display windows. The location of a peak, "PeakLoc," and its amplitude, "Amplitu," can be determined. The values of PeakLoc and Amplitu change as the cursor moves along.

  With NUTSA, it is intended that function keys and commands be employed to locate and characterize changes. A simple algorithm for quantifying the difference in width of waveform spikes was incorporated to allow for a gross quantification of change. As can be seen in Figure 11-12, by moving the cursor to the location of the spike difference apex, which is located at 30.12 ft, not only the excursion amplitude of –15.00 mρ is displayed at the left window, but also the width (PWidth) can be displayed by pressing the keys <Ctrl><D>.

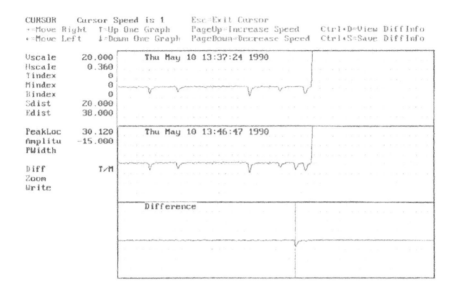

Figure 11-12. Using the cursor, it is possible to quantify the waveform magnitude and location. By pressing <Ctrl><D>, the peak width will be displayed.

This computational assistance is the first step in developing an expert system or an artificial intelligence program for analyzing changes in TDR signatures. It not only facilitates data interpretation but also facilitates quantifying and synthesizing of large volumes of acquired data. More sophisticated methods need to be developed to interpret large databases of TDR signatures that result from either long or multiple cables. Unfortunately, the large variation in shapes requires a significantly complex expert system to replace visual inspection by a human expert.

The "Diff," "Zoom," and <Ctrl><D> features of NUTSA were motivated by techniques currently employed to interpret TDR signature changes. For example, water level changes in Figure 11-13 were identified using "Diff" and "Zoom." The times and corresponding locations of water levels are summarized in Table 11.7. If a large number of water level changes are recorded, NUTSA provides an easy way to identify the water levels by

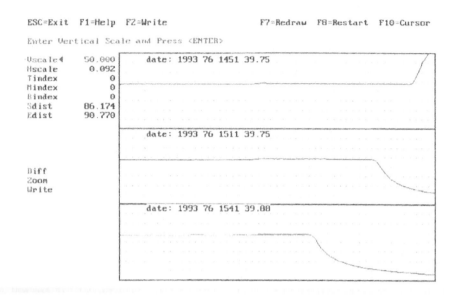

Figure 11-13. Waveforms obtained from the air-dielectric cable in Figure 11-2 as the water level was raised.

Table 11.7. Summary of TDR signature changes for water level example in Figure 11-13.

Time	Location of water (m)
14:51	90.421
15:11	89.851
15:41	88.923

locating the reflection created at the air–water interface. This technique can be applied to monitor water levels in wells, water levels in a reservoir, etc.

When the "Diff" command is invoked, NUTSA computes

the arithmetic difference between data points of two waveforms. The example in Table 11.8 illustrates how the difference waveform is computed using the following algorithm:

> *Adjusted byte value*
> > $= original\ byte\ value\ \text{-}\ vertical\ offset$ (11-1a)

where,

> *vertical offset for CSI1.DAT*
> > $= first\ data\ in\ block\ 2\ \text{-}\ last\ data\ in\ block\ 1$
> > $= 2715\ \text{-}\ 4216 = \text{-}1501,$ (11-1b)

and

> *vertical offset for CSI2.DAT*
> > $= first\ data\ in\ block\ 2\ \text{-}\ last\ data\ in\ block\ 1$
> > $= 5767\ \text{-}\ 4225 = 1542.$ (11-1c)

So that,

> *Byte value difference*
> > $= (byte\ value)_{CSI2.DAT}\ \text{-}\ (byte\ value)_{CSI1.DAT}.$ (11-2)

> *Difference graphical amplitude*
> > $= byte\ value\ difference,$ (11-3)

and

> *Reflection coefficient difference amplitude*
> > $= diff.\ graph.\ ampl.\ /\ 875\ graphical\ units/div$
> > > $x\ gain.$ (11-4)

Storing waveforms as ASCII files

The raw waveform data files shown in Tables 11.4 and 11.5 are simply the waveform values in some arbitrarily selected graphical scale. When comparing waveforms or attempting more sophisticated analysis, it is necessary to ensure that common bases are employed. This is one of the strengths of programs such as NUTSA, since values are computed for each data point along the entire length of a transducer cable or probe. These values are stored in ASCII format for use with any spreadsheet or plotting program.

Table 11.8. Calculation of difference waveform data points.

Data point	Peak Loc (m)	Byte value				Difference		
		Raw file CSI1.DAT	Adjusted file CSI1.DAT	Raw file CSI2.DAT	Adjusted file CSI2.DAT	Diff.	Graph. units	Reflect. coeff. (mrho)
140	89.750	2693	4194	5748	4206	12	12	0.686
141	89.750	2695	4196	5742	4200	4	4	0.229
142	89.768	2693	4194	5740	4198	4	4	0.229
143	89.777	2692	4193	5742	4200	7	7	0.400
144	89.787	2691	4192	5743	4201	9	9	0.514
145	89.796	2693	4194	5742	4200	6	6	0.343
146	89.805	2693	4194	5742	4200	6	6	0.343
147	89.814	2690	4191	5745	4203	12	12	0.686
148	89.823	2692	4193	5737	4195	2	2	0.114
149	89.833	2694	4195	5734	4192	-3	-3	-0.171
150	89.842	2688	4189	5735	4193	4	4	0.229
151	89.851	2689	4190	5731	4189	-1	-1	-0.057
152	89.860	2689	4190	5721	4179	-11	-11	-0.629
153	89.869	2690	4191	5709	4167	-24	-24	-1.371
154	89.879	2688	4189	5697	4155	-34	-34	-1.943
155	89.888	2687	4188	5677	4135	-53	-53	-3.029
156	89.897	2691	4192	5647	4105	-87	-87	-4.971
157	89.906	2693	4194	5614	4072	-122	-122	-6.971
158	89.915	2692	4193	5562	4020	-173	-173	-9.886
159	89.925	2692	4193	5503	3961	-232	-232	-13.257
160	89.934	2690	4191	5433	3891	-300	-300	-17.143
161	89.943	2690	4191	5360	3818	-373	-373	-21.314
162	89.952	2689	4190	5276	3734	-456	-456	-26.057
163	89.961	2689	4190	5191	3649	-541	-541	-30.914

Requirements and limitations

NUTSA was written using Microsoft QuickBASIC version 4.5. It requires an IBM-compatible computer using MS-DOS 3.3 or higher with a minimum 520K memory and a VGA, EGA, Hercules, or CGA display monitor. A hard drive is recommended to allow for analysis of large files, since scratch files must be created.

The number of data blocks (i.e., screens) required to store waveforms for an entire cable will depend on the horizontal scale used when the waveforms are acquired. If a horizontal sensitivity of 2 ft/div is used, then 5 data blocks are required for a 100 ft cable and 50 data blocks for a 1000 ft cable. The maximum number of data points that can be read into the NUTSA program for display is 10001 (with 39 data points overlapped). Therefore, the maximum number of data blocks that can be displayed is

$$(10001 + 39 \, points) / 251 \, points \, per \, data \, block$$
$$= 40 \, data \, blocks. \qquad (11\text{-}5)$$

If the data file is larger than 40 data blocks, it will not be possible to view the entire waveform. However, the waveform can be viewed in portions by changing Sdist and Edist. NUTSA will read scratch files and search for new Sdist points.

NUTSA was developed primarily for visual analysis of signature changes. It is intended that the <Diff>, <Zoom>, and cursor movement features will be used to locate and characterize changes. A very simple algorithm for quantifying the width of difference waveform spikes was incorporated to allow for a gross quantification of changes in width. More sophisticated versions are presented in the literature (Tektronix, 1989b; Baker and Allmaras, 1990; Heimovaara and Bouten, 1990; Reeves and Elgewazi, 1992). Software development is being pursued by many users of TDR technology. The number of field applications is increasing; therefore, the need for batch processing of TDR records will become more critical.

APPENDIX A
CABLE-GROUT PROPERTIES

CABLE SHEAR PROBES

There is a wide variety of commercially available cables that can be employed directly for TDR measurement of rock and soil shearing. In order to discuss how the selection of coaxial cable will affect the performance of a TDR monitoring system, it is necessary to first consider features of the various types of semiflexible cables that can be employed in a TDR monitoring system. For purposes of this discussion, coaxial cables are categorized into three main types: 1) smooth-wall aluminum outer conductor, 2) corrugated copper outer conductor, and 3) braided outer conductor. All three types of cables are available with either a polyethylene- or air-dielectric between the inner and outer conductor.

Smooth-wall aluminum coaxial cables (Figure 10-4a and Table 10.1) have an annular conductor which is fixed and uniform in geometry. Corrugated cables (Figure 10-4b) consist of an annular corrugated outer conductor for increased flexibility and easier bending as compared with the smooth-wall aluminum cable. Braided cables (Figure 10-4c) are even more flexible than those with smooth walls. Common cable diameters are 9.5 mm, 12.5 mm, 19 mm, and 22.2 mm (i.e., 3/8-in., 1/2-in., 3/4-in., and 7/8-in.). Air-dielectric cable is also available and is less "lossy" (results in less signal attenuation) compared with polyethylene dielectric and is used with some modification for measuring water level.

Engineering and electrical properties for various coaxial cables are given in Table 10.1. Tensile strength was estimated either based on the results of laboratory tests or by multiplying the cross-sectional area of the outer conductor times the tensile strength of aluminum (83 MPa or 12,000 psi) or copper (206 MPa or 30,000 psi). The capacitance is either based on values provided by the manufacturer or computed with Equation 2-21. The cables listed in Table 10.1 have been deployed in projects discussed in Chapters 5 through 9. While one of the main considerations has been cost, other factors have overridden cost. For example, the outer corrugated conductor is more sensitive to extension than shear deformation. Experience has shown that the electrical properties that have the greatest impact on the quality of TDR reflection waveforms are 1) resistance and 2) attenuation. Both of these factors control the magnitude of TDR reflection for a given magnitude of cable deformation. Braided cable has not been extensively deployed for fear that the braids will redistribute themselves with time and thus degrade the reflection detected by TDR, although it appears that they may provide the optimal response for monitoring localized shear in soil.

Experience has shown that the mechanical cable properties which have the greatest impact on TDR monitoring of rock and soil deformation are 1) solid aluminum outer conductor, 2) minimum bending radius, 3) tensile strength, 4) diameter of outer conductor, and 5) dielectric type. Using a solid aluminum outer conductor not only makes it possible to create permanent reference crimps, but also greatly enhances sensitivity to shear deformation. The range of movement that can be detected before a cable is short-circuited depends on the cable diameter.

In order to maximize the bond between the outer conductor and grout, cables should be ordered without an outer plastic jacket, if possible. In some situations, such as the *in situ* mine discussed in Chapter 6, the downhole environment is corrosive and a jacketed cable should be specified.

INSTALLATION AND GROUTING RECORDS

As the saying goes, "when in grout...doubt." The most tenuous and most critical component of a successful installation is proper and complete placement of the grout that bonds the cable to the surrounding rock mass. There are several excellent books that explain grouting procedures in the construction and oil industries (Weaver, 1991; Xanthakos et al., 1994; Jefferis, 1982; Houlsby, 1990), and one or two of these should be with the installation technician to complement the driller's wealth of experience. For each cable installed, it is very important to keep a record of installation details, including

> project name
> hole location and designation
> drilling dates
> hole conditions, including water levels
> cable crimp locations
> date when cable placed in hole
> length of cable in hole
> date of grouting
> technique used for tremie placement of grout
> grouting pipe type and diameter
> records of grout mix
> records of grout take versus estimated hole volume
> samples of grout
> testing of grout samples

As demonstrated by this list, there are several important features of a successful installation similar to any type of borehole instrumentation, not the least of which is proper mixing and placement of the grout. While it is tempting to simply mix grout and pump it downhole without making notes about grout mix and volume, discipline and documentation are required to ensure successful grout placement (although mother nature may have other ideas in mind when rock is highly fractured). Among the factors which must be addressed for every cable are 1) maintenance of a stable hole, even if this requires installing the cable through a PVC pipe which is left in place, 2) expansion agent, 3) additives to fill fractures and prevent grout loss, and 4) downhole plugs.

GROUT MIX DESIGN

Experience has shown that the grout properties that have the greatest impact on installation success and sensitivity to rock deformation are 1) viscosity, 2) solids content and fluid density, 3) expansion and minimization of voids, 4) compressive strength, 5) stiffness, 6) brittle fracture, and 7) bond strength and reaction between the grout and aluminum outer conductor. Grout property tests have been conducted to design a sufficiently strong expansive grout mixture that can be placed with a typical drilling rig water pump. The grout mix must have a low enough viscosity to allow pumping without sacrificing strength. At the same time, it must have the properties to prevent shrinkage to ensure that the grout will transfer rock deformation to the cable. These properties have been quantified through tests for viscosity, cable-grout bond strength, shrinkage, and unconfined compressive strength for various ratios of water, cement, and additive. Additives are employed to 1) make the grout flow more easily through pumps and 2) limit shrinkage.

Typically, grout mixtures are prepared in a 455 to 910 l (100 to 200 gallon) tub which drillers carry on their drill truck. The driller will pump 225 to 455 l (50 to 100 gallons) of water into the tub, then use the pump to recirculate it as bags of cement and additive are emptied into the water. When the pump is working too hard, he will not add more cement. A brief summary of grout mixes that have been successfully pumped and cleaned are listed in Table A.1. The following table translates between water:cement (w/c) ratios and bags cement:gal water ratios:

w/c (%)	bags cement per 100 gal water (no. of 94-lb bags)	bags cement per 1000 liter water (no. of 25-kg bags)
50	17	80
60	14	70
70	12	60
80	10	50
100	8	40
120	7	30
200	4	20
400	2	10

where 94 lb = 43 kg, 100 gal = 455 liter, and water weighs 1 kg/liter. Field experience shows that drillers will hesitate to add more than 12 (94-lb) bags to 100 gal of water due to concerns

Table A.1. Grout mixes which have been used at TDR installations.

Location	Portland cement type	Bentonite W_{ben}/W_c %	Additive[1] W_{add}/W_c %	Water W_w/W_c %
San Manuel, AZ	n/a			51
Pecos, TX	n/a			60
Oakwood, VA	Type I			53
Collinsville, IL	Type I		2.7	65
West Frankfort, IL	Type I		1.9	91
Benton, IL	Type I		0.8	72
Collinsville, IL	Type I		1.1	97
	Type I		0.9	63
	Type I		1.3	76
	Type I		1.8	85
	Type I		1.6	68
Ft. McMurray, Alberta	Type 10	7.1		60
Orangeville, UT	Type I-II	4.4		71
Horse Creek, CA	Type I-II		2.6	85
Oakland, MD	Type II			128
Mt. Storm, WV	Type II			128
Marion, IL	Type III		2.7	65
Sesser, IL	Type III		2.7	65
			min	51
			max	128
			avg	77

[1] Intrusion-Aid Type R, Specrete-IP Inc., Cleveland, OH.

about damaging pumps and the pressure required to pump the mixture through a 25 mm (1-in.) grout pipe. This concern is reflected by the frequency of w/c between 60% and 80% in Table A.1.

Su (1987) determined the viscosity of typical fluid grouts with a Brookfield Viscometer by measuring the force required to rotate a spindle in the grout at a given angular velocity. Increasing the water:cement ratio from 40% to 60% (i.e., reducing the number of 94 lb bags of cement per 100 gal of water from 21 to 14) significantly decreased the viscosity (from 600 centipoise to 60 centipoise), but increasing this ratio beyond 60% had little effect. When different amounts of Intrusion-Aid agent were added to Type III Portland cement with a water:cement ratio of 65%, it was found that increasing the amount of additive beyond 1% (i.e., 1 lb of additive per 94 lb bag of cement) had little effect on reducing viscosity.

Strength and shrinkage properties of the grout must be such that an adequate bond develops between the grout and cable and between the grout and borehole wall. Therefore, as shown in Table A.2, the unconfined compressive strength, adhesive strength, and volume change characteristics have been determined for a variety of grout mixtures prepared in the laboratory and the field.

The adhesive strength between grout and unjacketed aluminum outer conductor cable was determined by performing either 1) pull-out tests on a section of cable grouted into molds, or 2) extension tests with cable grouted into steel pipe. For all tests, the molds had an inside diameter of 50 mm. Su (1987) performed tests using 9.5, 12.7, and 22.2 mm Cablewave cable. Peterson (1993) performed tests using 12.7 and 19 mm CommScope cable and 12.7 and 22.2 mm Cablewave cable. The results are summarized in Table A.3, including data from Mindess and Young (1981) for pull-out tests done with a plain steel bar in cement paste mixed at water:cement ratio of 80%. It can be seen that the results depend on 1) embedded length, 2) grout composition, and 3) cable or rod material and stiffness. See texts on reinforced concrete behavior for a discussion of development length along reinforcing bars (e.g., Park and Paulay, 1975).

Since it is important that unjacketed cable be used, complications arise due to a weak zone which develops due to the chemical reaction between the high pH cement and aluminum (or steel) as discussed in Mindess and Young (1981). Aluminum mixed with aqueous cement produces sulfur dioxide gas. These gas bubbles produces a much less dense zone immediately adjacent to the cable (Kim, 1989; Su, 1987). The compressive strength of this zone is noticeably low compared to grout a greater distance from the cable. This weak zone has been found to affect the localization of cable deformation when the grout fractures, as discussed in

Table A.2. Engineering properties of grout samples.

	Cement	Water W_w/W_c wt%	Additive W_{add}/W_c wt%	Bentonite W_{bent}/W_c wt%	Unconfined comp. strength MPa	Young's modulus GPa	Poisson's ratio	Volume Change†† %
				Field Samples				
Oakwood, VA	Type I	53			35.9	6.74	0.18	
		53			24.5	8.51	0.19	
Orangeville, UT	Type I-II	71		4.4	40.1	6.85	0.17	
		71		4.4	34.7	6.35	0.17	
Horse Creek, CA	Type I-II	85	2.6*		4.8			+3
				Laboratory Samples				
Su (1987)	Type III	40	1.0*		12.8			+2
		50	1.0*		11.3			+1
		55	0.5*		22.2			–3
		60	1.0*		9.9			–2
		65			1.4			–14
		65	1.0*		8.6			–8
		65	2.0*		8.5			–16
		75	0.5*		7.5			–17
O'Connor (1991)	Type 10	59		7.1†	22.0			
Peterson (1993)		59		7.1†	20.0			
	TEL Master	59	2.0‡		11.0			
		54	2.0‡		14.0			

* Intrusion-Aid Type R, Specrete-IP Inc., Cleveland, OH　　† Quick Gel, Baroid Ltd, Onoway, Alberta
‡ Ca, additive, TEL = thixotropic, expansive, low fluid loss, Canadian Fracmaster Ltd, Calgary, Alberta
†† + = expansion, – = shrinkage

Table A.3. Summary of laboratory adhesion strength tests.

Test type	Cement	Water W_w/W_c wt%	Additive W_{add}/W_c wt%	Bentonite W_{bent}/W_c wt%	Embedded length of cable mm	Adhesive strength MPa
pull-out, smooth aluminum Su (1987)	Type III	40	1.0*		89	0.49
		50	1.0*		89	0.45
		55	0.5*		89	0.37
		55	0.5*		89	0.28
		60	1.0*		89	0.39
		65			89	0.27
		65	1.0*		89	0.22
		65	2.0*		89	0.21
		75	0.5*		89	0.21
pull-out, smooth aluminum Peterson (1993)	Type 10	59		7.1†	85	0.04
		59		7.1†	85	0.18▲
		59			85	0.15
	TEL Master	65	2.0‡		85	0.01
		59	2.0‡		85	0.16
		54	2.0‡		85	0.04
pull-out, steel bar Mindess and Young (1981)		80				0.90

* Intrusion-Aid Type R, Specrete-IP Inc., Cleveland, OH   † Quick Gel, Baroid Ltd., Onoway, Alberta   ▲ cable coated

‡ Ca, additive, TEL = thixotropic, expansive, low fluid loss, Canadian Fracmaster Ltd., Calgary, Alberta

Chapter 5. Results from a series of tests done by Peterson (1993) show that coating the cable increased the relative adhesive strength from 0.04 to 0.18 MPa, as shown in Table A.3. This problem is remedied by spray painting the bare aluminum cable.

APPENDIX B
REFERENCES

Aimone-Martin, C.T. and Francke, J.L., Time Domain Reflectometry (TDR): A Comparison of Field Data to Laboratory Shear Tests, *Int. J. Rock Mech. & Min. Sci.* Vol. 34, No. 3-4, Paper No. 084, 1997, 12 pp.

Aimone-Martin, C.T., Oravecz, K.I., and Nytra, T.K., TDR Calibration for Quantifying Rock Mass Deformation at the WIPP Site, Carlsbad, New Mexico, *Proceedings of the Symposium on Time Domain Reflectometry in Environmental, Infrastructure, and Mining Applications*, Evanston, Illinois, Sept 7-9, U.S. Bureau of Mines, Special Publication SP 19-94, 1994, NTIS PB95-105789, pp. 507-517.

Alharthi, A. and Lange, J., Soil Water Saturation: Dielectric Determination, *Water Resour. Res.*, Vol. 23, 1987, pp. 591-595.

Allard, F.C., *Fiber-Optic Handbook for Engineers and Scientists*, McGraw-Hill Publishing Co., New York, 1990.

Alvenäs, G. and Stenberg, M., Problems in Estimating Soil Water Content by TDR Measurements, *Proceedings of the Symposium: Time-Domain Reflectometry Applications in Soil Science*, (Foulum, Denmark, Sept 16, 1994) SP Report No. 11, June, Vol. 3, 1995, pp. 121-123.

American Institute of Physics Handbook, 2nd ed., D. Gray, Ed., McGraw-Hill Book Co., New York, 1957, pp. 48-49.

American Society for Testing and Materials, *Standard Practice for Design and Installation of Ground Water Monitoring Wells in Aquifers*, D5092-90, Philadelphia, V.04.08, 1990, pp. 1-12.

Andrews, J.R., Time Domain Reflectometry, *Proceedings of the Symposium on Time Domain Reflectometry in Environmental, Infrastructure, and Mining Applications, Evanston, Illinois*, Sept 7-9, U.S. Bureau of Mines, Special Publication SP 19-94, 1994, NTIS PB95-105789, pp. 4-13.

Annan, A.P., Time Domain Reflectometry - Air-Gap Problem in a Coaxial Line, Report of Activities, Part B, Paper 77-1B, Geol. Surv. Canada, Ottawa, 1977, pp. 55-58.

Ansari, F., Editor, *Applications of Fiber Optic Sensors in Engineering Mechanics*, Am. Soc. of Civil Engineers, New York, 1993, 336 pp.

Askins, G.C., Putman, M.A., Williams, G.M., and Friebele, E.J., Stepped-Wavelength Optical Fiber Bragg Grating Arrays Fabricated in Line on a Draw Tower, *Optics Lett.*, Vol. 19, 1994, p. 147.

Aston, T. and Charette, F., Installation and Monitoring of Three Abandoned Mine Crown Pillar Sites, Timmons, Ontario: August 1992 to October 1992, CANMET Div. Rep. MRL 93-020 (CL), Min. Res. Lab., Ottawa, June, 1993, 61 pp.

Bailey, D.S., Corrosion Resistant Cable, U.S. Patent 5,355,720, October 18, 1994.

Baker, J.M. and Allmaras, R.R., System for Automating and Multiplexing Soil Moisture Measurement by Time Domain Reflectometry, *Soil Sci. Soc. Am. J.*, Vol. 54, No. 1, 1990, pp. 1-6.

Baker, J.M. and Lascano, R.J., The Spatial Sensitivity of Time Domain Reflectometry, *Soil Science*, Vol 14, No. 5, 1989, pp. 378-384.

Baker, T.H.W., Davis, J.L., Hayhoe, H.N., and Topp, G.C., Locating the Frozen-Unfrozen Interface in Soils Using Time Domain Reflectometry, *Can. Geotech. Jour.*, Vol. 19, 1982, pp. 511-517.

Balanis, C.A., Jeffrey, J.L., and Yoon, Y.K., Electrical Properties of Eastern Bituminous Coal as a Function of Frequency, Polarization and Direction of Electromagnetic-Wave and Temperature of Sample, *IEEE Trans. Geosc. Electronics*, GE16, 4, 1978, pp. 316-323.

Baran, E., Use of Time Domain Reflectometry for Monitoring Moisture Changes in Crushed Rock Pavements, *Proceedings of the Symposium on Time Domain Reflectometry in Environmental, Infrastructure, and Mining Applications*, Evanston, Illinois, Sept 7-9, U.S. Bureau of Mines, Special Publication SP 19-94, 1994, NTIS

345

PB95-105789, pp. 349-356.

Barlo, T.J., Dowding, C.H., and Prine, D.W., Detection of Leaking Liquids Through Measurement by Time Domain Reflectometry, Disclosure of Invention, 1996.

Bartel, E. W., Fox, M., Borgoyn, E., and Wingfield, P., TDRM Testing (contract J0377021), BuMines OFR 60-82, 1980, 135 pp.

Bauer, R.A., Dowding, C.H., Van Roosendaal, D.J., Mehnert, B.B., Su, M.B., and O'Connor, K., *Application of Time Domain Reflectometry to Subsidence Monitoring*, Office of Surface Mining, Pittsburgh, 1991, 48 pp.; NTIS PB 91-228411.

Bennett, R.M., Drumm, E.C., and Lin, G., Evaluation and Analysis of Remedial Measures for Earth-Structure Damage Mitigation, Final report, U.S. Bureau of Mines Contract No. SO278058, University of Tennessee, 1992.

Benson, C.H., Bosscher, P.J., Lang, D.T., and Pliska, R.J., Monitoring System for Hydrolgic Evaluation of Landfill Covers, *Geotechnical Testing Journal*, GTJODJ, Vol. 17, No. 2, June, 1994, pp. 138-149.

Benson, C.H., Using Time Domain Reflectometry (TDR) to Measure Volumetric Water Content, Course notes, University of Wisconsin-Madison, July, 1997.

Bjerrum, L., Progressive Failure in Slopes of Overconsolidated Plastic Clay and Clay Shales, *Proceedings*, American Society of Civil Engineers, Vol. 93, No. SM5, 1967, pp. 3-49.

Boehm, R.G. and Scherbert, G.S., Development of the Glendale Landfill Test Cover Study, *Proceedings, Waste Tech '97*, Tempe, Arizona, February, 1997, pp. 95-120.

Bohl, H. and Roth, K., Evaluation of Dielectric Mixing Models to Describe the $\theta(\epsilon)$ Relation, *Proceedings of the Symposium on Time Domain Reflectometry in Environmental, Infrastructure, and Mining Applications*, Evanston, Illinois, Sept 7-9, U.S. Bureau of Mines, Special Publication SP 19-94, 1994, NTIS PB95-105789, pp. 309-319.

Brigham, E.O., *The Fast Fourier Transform*, Prentice-Hall, Englewood Cliffs, NJ, 1980.

Brutcher, D.F., Mehnert, B.B., Van Roosendaal, D.J., and Bauer, R.A., Rock Strength and Overburden Changes Due to Subsidence Over a Longwall Coal Mining Operation in Illinois, *Paper in Rock Mechanics Contributions and Challenges* (Proc. 31st U.S. Sym. on Rock Mech., Golden, CO), Balkema, 1990, pp. 563-570.

Burland, J.B. and Moore, J.F.A., The Measurement of Ground Displacement Around Deep Excavations, *Proceedings, Symposium on Field Instrumentation*, British Geotechnical Society, London, 1973, pp. 70-84.

Burland, J.B., Longworth, T.I., and Moore, J.F.A., A Study of Ground Movement and Progressive Failure Caused by a Deep Excavation in Oxford Clay, *Geotechnique*, v. 27, No. 4, 1977, pp. 557-591.

Cablewave Systems (North Haven, CT), *Antenna and Transmission Line Systems*, Catalog 600, 1985, 208 pp.

Campbell, G.S. and Campbell, M.D., Irrigation Scheduling Using Soil Moisture Measurements: Theory and Practice, *Advances in Irrigation*, D. Hillel, Ed., Vol. 1, Academic Press, New York, 1982, pp. 25-42.

Campbell Scientific, Inc. (Logan, UT), *Time Domain Reflectometry for Measurement of Rock Mass Deformation*, Product brochure, July, 1991, 2 pp.

Campbell Scientific, Inc. (Logan, UT), *TDR Soil Moisture Measurement User's Manual*, 1992a.

Campbell Scientific, Inc. (Logan, UT), *PC208 Datalogger Support Software Instruction Manual*, 1992b.

Campbell Scientific, Inc. (Logan, UT), *CR10 User's Manual*, 1992c.

Cedergren, H.R., *Seepage, Drainage, and Flow Nets*, John Wiley & Sons, New York, 1967.

Charette, F., Installation and Monitoring of Three Abandoned Mine Crown Pillar Sites, Cobalt, Ontario: February 1992 to July 1992, CANMET Div. Rep. MRL 92-095(CL), Min. Res. Lab., Ottawa, Mar., 1993a, 64 pp.

Charette, F., Results of the Monitoring of Three Crown Pillar Sites in Cobalt, Ontario, CANMET Div. Rep. 92-101 (CL), Min. Res. Lab., Ottawa, Mar., 1993b, 31 pp.

Cole, R.H., Time Domain Spectroscopy of Dielectric Materials, *IEEE Transactions I&M*, Vol. IM25, No. 4, Dec., 1976, pp. 371-375.

CommScope, Inc. of North Carolina (Claremont, NC), *Cable Products Catalog*, 1994.

CSIRO (Canberra), *PyeLab TDR System - User's Guide Version 3.205*, CSIRO Centre for Environmental Mechanics, 1991.

Dalton, F.N., Development of Time Domain Reflectometry for Measuring Soil Water Content and Bulk Soil Elevtrical Conductivity, *Advances in Measuremnt of Soil Physical Properties: Bringing Theory into Practice*, G. C. Topp et al., Eds., Soil Sci. Soc. Am., Madison, WI, SSSA Sp. Pub. 30, 1992. pp. 143-167.

Dalton, F.N., and van Genuchten, M.T., The Time Domain Reflectometry Method for Measuring Soil Water Content and Salinity, *Geoderma*, 38, 1986, pp. 237-250.

Dalton, F.N., Herkelrath, W.N., Rawlins, D.S., and Rhoades, J.D., Time Domain Reflectometry: Simultaneous Measurement of Soil Water Content and Electrical Conductivity with a Single Probe, *Science*, Vol. 224, June,1984, pp. 989-990.

Dames and Moore, A Demonstration of Longwall Mining— Appendix (contract J0333949, Old Ben Coal Co.), BuMines OFR 86(2)-85,1983, 371 pp.

Dandridge, A., Cole, J.H., and Siegel, Jr., G.H., Optical Fiber Sensors, *Technical Digest - Symposium on Optical Fiber Measurements*, 1984, pp. 49-54.

Dasberg, S. and Dalton, F.N., Time Domain Reflectometry Field Measurements of Soil Water Content and Electrical Conductivity, *Soil Sci. Soc. Am. J.*, Vol. 49, 1985, pp. 293-297.

Dasberg, S. and Hopmans, J.W., Time Domain Reflectometry Calibration for Uniformly and Nonuniformly Wetted Sandy and Clayey Loams, *Soil Sci. Soc. Am. J.*, Vol. 56, 1992, pp. 1341-1345.

Davis, C.M., Carome, E.F., Weik, M.H., Ezekiel, S., and Einzig, R.E., *Fiberoptic Sensor Technology Handbook*, Optical Technologies (OPTECH), Herndon, VA, 1986, 107 pp.

Davis, J.L. and Chudobiak, W.J., In-situ Meter for Measuring Relative Permittivity of Soils, Pap 75-1A, Geol. Surv. Can., Ottawa, 1975, pp. 75-79.

De Clerck, P., Mesure de l'Evolution de la Teneur en Eau des par Voie Electromagnetique, *Tech. Routiere*, Vol 3, 1985, pp. 6-15.

De Loor, G.P., Dielectric Properties of Heterogeneous Mixtures Containing Water, *J. Microwave Power*, Vol. 3-2, 1968, pp. 67-73.

Dirksen, C.E. and Dasberg, S., Four Component Mixing Model for Improved Calibration of TDR Soil Water Content Measurements, *Soil Sci. Soc. Amer. J.*, Vol. 57, 1993, pp. 660-667.

Dobson, M.C., Ulaby, F.T., Hallikainen, M.T., and El-Rayes, M.A., Microwave Dielectric Behavior of Wet Soil, II. Dielectric Mixing Models, *IEEE Trans. Geosci. Remote Sens.*, Vol. GE-23, No. 1, 1985, pp. 35-46.

Doepken, W.G., Climax Completing $40 Million Modernization of Henderson Mine and Mill, *Min. Eng.*, Nov.,1991, pp. 1315-1316.

Dowding, C.H., Internal report, Dept. of Civil Eng., Northwestern University, Evanston, IL, 1983.

Dowding, C.H. and Huang, F.-C., Telemetric Monitoring for Early Detection of Rock Movement With Time Domain Reflectometry, *J. Geot. Eng., Am. Soc. Civ. Eng.*, V. 120, No. 8, 1994a, pp. 1413-1427.

Dowding, C.H. and Huang, F.-C., Ground Water Pressure Measurement with Time Domain Reflectometry, *Proceedings of the Symposium on Time Domain Reflectometry in Environmental, Infrastructure, and Mining Applications*, Evanston, Illinois, Sept 7-9, U.S. Bureau of Mines, Special Publication SP 19-94, 1994b, NTIS PB95-105789, pp. 247-258.

Dowding, C.H., Huang, F.-C., and McComb, P.S., Groundwater Pressure Measurement with Time Domain Reflectometry, *Geotechnical Testing Journal*, GTJODJ, ASTM, Vol. 19, No. 1, March, 1996, pp. 58-64.

Dowding, C.H. and Pierce, C.E., Measurement of Localized Failure Planes in Soil with Time Domain Reflectometry, *Proceedings of the Symposium on Time Domain Reflectometry in Environmental, Infrastructure, and Mining Applications*, Evanston, Illinois, Sept 7-9, U.S. Bureau of Mines, Special Publication SP 19-94, 1994a, NTIS PB95-105789, pp. 569-578.

Dowding, C.H. and Pierce, C.E., Use of Time Domain Reflectometry to Detect Bridge Scour and Monitor Pier Movement, *Proceedings of the Symposium on Time Domain Reflectometry in Environmental, Infrastructure, and Mining Applications*, Evanston, Illinois, Sept 7-9, U.S. Bureau of Mines, Special Publication SP 19-94, 1994b, NTIS PB95-105789, pp. 579-587.

Dowding, C.H., Pierce, C.E., Nicholson, G.A., Taylor, P.A., and Agoston, A., Recent Advancements in TDR Monitoring of Ground Water Levels and Piezometric Pressures, *Proceedings of the 2nd North American Rock Mechanics Symposium: NARMS'96*, Montreal, June, 1996, pp. 1375-1381.

Dowding, C.H., Su, M.B., and O'Connor, K.M., Principles of Time Domain Reflectometry Applied to Measurement of Rock Mass Deformation, *Int. J. Rock Mech. and Min. Sci.*, v. 25, No. 5, 1988, pp. 287-297.

Dowding, C. H., Su, M. B., and O'Connor, K. M., Measurement of Rock Mass Deformation with Grouted Coaxial Antenna Cables, *Rock Mechanics and Rock Engineering*, Vol. 22, 1989, pp. 1-23.

Drungil, C.E.C., Abt, K., and Gish, T.J., Soil Moisture Determination in Gravelly Soils with Time Domain Reflectometry, *Trans. Am. Soc. Agr. Eng.*, Vol. 32, 1989, pp. 177-180.

Dubaniewicz, T.H., Heasley, K.A., and DiMartino, M.D., Fiber-Optic Sensor Development, *Proceedings, Symposium on Time Domain Reflectometry in Environmental, Infrastructure, and Mining Applications*, Evanston, Illinois, Sept 7-9, U.S. Bureau of Mines, Special Publication SP 19-94, 1994, NTIS PB95-105789, pp. 461-471.

Dunicliff, J., *Geotechnical Instrumentation for Monitoring Field Performance*, John Wiley, New York, 1988.

Dworak, R.A., Jordan, A.G., and Thorne, J.S., Time Domain Reflectometer Microcomputer (contract H0346138), BuMines OFR 106-77, 1977, 113 pp.

Dworsky, L.N., *Modern Transmission Line Theory and Application*, John Wiley & Sons, New York, 1979.

Easy Test, Ltd. (Lublin, Poland), *EASY TEST Soil Water Status Monitoring Devices*, 1994.

Elias, V. and Juran. I., Soil Nailing for Stabilization of Highway Slopes and Excavations. U.S. Federal Highway Administration, Pub. No. FHWA-RD-89-193, June, 1991.

Environmental Protection Agency, *Procedures Manual for Ground Water Monitoring of Solid Waste Disposal Facilities*, Vol. 530/SW-G11, Aug., 1977, pp. 1-101.

ESI Environmental Sensors, Inc. (Victoria, BC), *MoisturePoint Model MP-917 Soil Moisture Measurement Instrument Operation Manual*, 1995.

Eveready Battery Co., Inc. (St. Louis, MO), *ENERGIZER Owners Manual*, 1995.

Fair, A.E. and Lord, E.R.F., Methods used to monitor and control block slides in oilsands at Syncrude's dragline operation in Northern Alberta, Canada, *Proceedings, 37th Can. Geotech. Conf.*, Toronto, Sept. 17,1984, pp. 103-112.

Fellner-Feldegg, J., The Measurement of Dielectrics in the Time Domain, *J. Phys. Chem.*, Vol. 73, 1969, pp. 616-623.

Finno, R.J., Atmatzidis, D.K., and Perkins, S.B., Observed Performance of a Deep Excavation in Clay, *Journal of Geotechnical Engineering*, ASCE, v. 115, No. 8, 1989a, pp. 1045-1064.

Finno, R.J. and Nerby, S.M., Saturated Clay Response During Braced Cut Construction, *Journal of Geotechnical Engineering*, ASCE, v. 115, No. 8, 1989b, pp. 1065-1084.

Francke, J.L. and Terrill, L.J., The Excavation Effects Program at the Waste Isolation Pilot Plant, *Proceedings, Int. Congress on Mine Design, Innovation in Mine Design for the 21st Century*, Kingston, Ontario, Aug., 1993, pp. 469-477.

Francke, J.L., Terrill, L.J., and Francke, C.T., Time Domain Reflectometry Study at the Waste Isolation Pilot Plant, *Proceedings of the Symposium on Time Domain Reflectometry in Environmental, Infrastructure, and Mining Applications*, Evanston, Illinois, Sept 7-9, U.S. Bureau of Mines, Special Publication SP 19-94, 1994, NTIS PB95-105789, pp. 555-568.

Freeman, E.L., Time Domain Reflectometry at Cloverdale Landslide, U.S. Highway 101, Sonoma County, California, Tech. Res. Rept. CE-96-03, Dept. of Civil Engineering, Univ. of the Pacific, April, 1996, 10 p.

Gardner, W.H., Water Content, *Methods of Soil Analysis*, Vol. 9, Part 1, 1986, pp. 493-523.

Gear, R.D., Dansfield, A.S., and Campbell, M.D., Irrigation Scheduling with Neutron Probe, *J. Irrig. Drain Div.*, Am. Soc. Civ. Eng., Vol. 103, 1977, pp. 291-298.

Guitini, J.C., Zanchetta, J.V., and Diaby, S., Characterization of Coals by the Study of Complex Permittivity, *Fuel*, Vol. 66, 1987, p. 179.

Guymon, G.L., Berg, R.L., and Hromadka, T.V., Mathematical Model of Frost Heave and Thaw Settlement in Pavements, U.S. Army Cold Regions Res. and Eng. Lab., Report N. 93-2, April, 1993.

Halbertsma, J., van den Elsen, E., Bohl, H., and Skierucha, W., Temperature effects on TDR determined soil water content, *Paper in the Proceedings of the Symposium: Time-Domain Reflectometry Applications in Soil Science*, (Foulum, Denmark, Sept 16, 1994) SP Report No. 11, June, Vol. 3, 1995, pp. 35-37.

Halliday, D. and Resnick, R., *Physics*, Part II, 2nd ed., John Wiley & Sons, New York, 1962.

Hasted, J.B., *Aqueous Dielectrics*, Chapman and Hall, London, 1973.

He, G. and Cuomo, F.W., Displacement Response, Detection Limit and Dynamic Range of Fiber-Optic Lever Sensors, *Journal of Lightwave Technology*, Vol. 9, No. 11, Nov., 1991, pp. 1618-1625.

Heimovaara, T.J., Frequency Domain Analysis of Time Domain
    Reflectometry Waveforms, 1. Measurement of the Complex
    Dielectric Permittivity of Soils, *Water Resour. Res.*, Vol.
    30, 1994, pp. 189-199.
Heimovaara, T.J. and Bouten. W., A Computer-Controlled 36-
    Channel Time Domain Reflectometry System for
    Monitoring Soil Water Contents, *Water Resources
    Research*, Vol. 26, No. 10, October, 1990, pp. 2311-2316.
Herkelrath, W.N., Hamburg, S.P., and Murphy, F., Automatic,
    Real-Time Monitoring of Soil Moisture in a Remote Field
    Area with Time Domain Reflectometry, *Water Resources
    Research*, Vol. 27, No. 5, 1991, pp. 857-864.
Hewlett-Packard (Palo Alto, CA), *TDR Fundamentals for Use with
    HP 54120T Digitizing Oscilloscope and TDR*, Application
    Note 62, April,1988, 57 pp.
Hilhorst, M. A. and Dirksen, C., Dielectric Water Content Sensors:
    Time Domain Versus Frequency Domain, *Proceedings of
    the Symposium on Time Domain Reflectometry in
    Environmental, Infrastructure, and Mining Applications*,
    Evanston, Illinois, Sept 7-9, U.S. Bureau of Mines, Special
    Publication SP 19-94, 1994, NTIS PB95-105789, pp. 23-
    33.
Hill, J.D., Monitoring of Surface Crown Pillar Deformation at
    Goldenville, N. S. using Time Domain Reflectometry, Final
    Report, DSS Contract No. 23440-0-9245/01-S2,
    CANMET, prep. by Dept. of Min. and Metall. Eng., Tech.
    Univ. Nova Scotia, Halifax, Mar., 1993, 94 pp.
Hill, K.O., Malo B., Bilodeau F., Johnson, D.C., and Albert, J.,
    Bragg Gratings Fabricated in Photosensitive Monomode
    Optical Fiber by UV Exposure Through a Phase Mask,
    Appl. Phy. Lett., Vol. 62, 1993, p. 1035.
Hokett, S.L., Russell C.E., and Gillespie, D.R., Water Level
    Detection During Drilling Using Time Domain
    Reflectometry, *Proceedings of the Symposium on Time
    Domain Reflectometry in Environmental, Infrastructure,
    and Mining Applications*, Evanston, Illinois, Sept 7-9, U.S.
    Bureau of Mines, Special Publication SP 19-94, 1994, NTIS
    PB95-105789, pp. 259-269.

Holmes, G.F. and List, B.R., Highwall Geotechnical Engineering at Syncrude, *International Journal of Surface Mining*, v. 3, 1989, pp. 21-26.

Hook, W.R. and Livingston, N.J., Errors in Converting Time Domain Reflectometry Measurements of Propagation Velocity to Estimates of Soil Water Content, *Soil Sci. Soc. Am. Journal*, Vol. 60, Jan.-Feb., 1996, pp. 35-41.

Hook, W.R., Livingston N. J., Sun, Z.J., and Hook, P.B., Remote Diode Shorting Improves Measurement of Soil Water by Time Domain Reflectometry, *Soil Sci. Soc. Am. J.*, Vol. 56, No. 5, 1992, pp. 1384-1391.

Horiguchi, T., Kurashima, T., and Koyamada, Y., Measurement of Temperature and Strain Distribution by Brillouin Frequency Shift in Silica Optical Fibers, SPIE 1797-01 in *Distributed and Multiplexed Fiber Optic Sensors II*, J. Dakin and A. Kersey, Eds., Sept. 1992, Boston, MA.

Houlsby, A.C., *Construction and Design of Cement Grouting*, John Wiley & Sons, New York, 1990.

Huang, F.-C., O'Connor, K.M., Yurchak, D.M., and Dowding, C.H., NUMOD and NUTSA: Software for Interactive Acquisition and Analysis of Time Domain Reflectometry Measurements, *BuMines IC 9346*, 1993, 42 pp.

Hubscher, R. and Or, D., *WinTDR Users Guide*, The USU Soil Physics Group, Utah State University, Logan, http://tal.agsci.usu.edu/home/ron/web/tdr/win_tdr.html, 1996.

Huston, D.R., Fuhr, P.L., and Ambrose, T.P., Damage Detection in Structures using OTDR and Intensity Measurements, *Proceedings of the Symposium on Time Domain Reflectometry in Environmental, Infrastructure and Mining Applications*, Evanston, Illinois, Sept. 7-9, U.S. Bureau of Mines, Special Publication SP19-94, 1994, NTIS PB95-105789, pp. 484-493.

IMKO GmbH (Ettlingen, Germany), *TRIME Product Guide*, 1995.

Iwata, S., Tabuchi, T., and Warkentin, B.P., *Soil-Water Interactions: Mechanisms and Applications*, Marcel Dekker, Inc., New York, 1988.

Jacobsen, O.H. and Schjønning, P., Comparison of TDR
    Calibration Functions for Soil Water Determination, *Paper
    in the Proceedings of the Symposium: Time-Domain
    Reflectometry Applications in Soil Science*, (Foulum,
    Denmark, Sept 16, 1994) SP Report No. 11, June, Vol. 3,
    1995, pp. 25-33.

Janoo, V., Berg, R.L., Simonsen, E., and Harrison, A., Seasonal
    Changes in Moisture Content in Airport and Highway
    Pavements, *Proceedings of the Symposium on Time
    Domain Reflectometry in Environmental, Infrastructure,
    and Mining Applications*, Evanston, Illinois, Sept 7-9, U.S.
    Bureau of Mines, Special Publication SP 19-94, 1994, NTIS
    PB95-105789, pp. 357-363.

Jefferis, S.A., Effects of Mixing on Bentonite Slurries and Grouts,
    *Proceedings, Conf. on Grouting in Geotechnical
    Engineering*, ASCE, New Orleans, 1982, pp. 62-76.

Johnson, B., Remote Monitoring of Piezometers with Automation,
    Proceedings, 44th Annual Geotech Eng. Conf., Univ. of
    Minnesota, Minneapolis, Feb., 1996, 58 pp.

Juran, I., Soil Nailing in Excavations, Notes prepared for the
    Cooperative Program, Project 24-2, C. B. Villet and J. K.
    Mitchell, Eds., Appnd. 3.A, Vol. VI, 1986.

Kane, W.F., Embankment Monitoring with Time Domain
    Reflectometry, in *Proc. Fifth Int. Conf. on Tailings and
    Mine Waste '98*, Fort Collins, 26-28 Jan., 1998a, pp. 223-
    230.

Kane, W.F., Personal communication, *KANE*GeoTech, Stockton,
    CA, 1998b.

Kane, W.F. and Beck, T.J., Rapid Slope Monitoring, *Civil
    Engineering*, Vol. 66, No. 6, June, 1996a, pp. 56-58.

Kane, W. F. and Beck, T. J., An Alternative Monitoring System for
    Unstable Slopes, *Geotechnical News*, Vol. 14, No. 3,
    September, 1996b, pp. 29-31.

Kane, W.F., Perez, H., and Anderson, N.O., Development of a
    Time Domain Reflectometry System to Monitor Landslide
    Activity, Tech. Report FHWA/CA/TL-96/09, Univ. of the
    Pacific, Stockton, June, 1996, 72 pp.

Kawamura, N., Bauer, R.A., Mehnert, B.B., and Van Roosendaal, D.J., TDR Cables, Inclinometers and Extensometers to Monitor Coal Mine Subsidence in Illinois, *Proceedings of the Symposium on Time Domain Reflectometry in Environmental, Infrastructure, and Mining Applications*, Evanston, Illinois, Sept 7-9, U.S. Bureau of Mines, Special Publication SP 19-94, 1994, NTIS PB95-105789, pp. 528-539.

Kaya, A., Lovell, C.W., and Altschaeffl, A.G., The Effective Use of Time Domain Reflectometry (TDR) in Geotechnical Engineering, *Proceedings of the Symposium on Time Domain Reflectometry in Environmental, Infrastructure, and Mining Applications*, Evanston, Illinois, Sept 7-9, U.S. Bureau of Mines, Special Publication SP 19-94, 1994, NTIS PB95-105789, pp. 398-409.

Kersey, A.D., Monitoring Structural Performance with Optical TDR Techniques, *Proceedings of the Symposium on Time Domain Reflectometry in Environmental, Infrastructure, and Mining Applications*, Evanston, Illinois, Sept 7-9, U.S. Bureau of Mines, Special Publication SP 19-94, 1994, NTIS PB95-105789, pp. 434-442.

Kim, M.H., Quantification of Rock Mass Movement with Grouted Coaxial Cables, M.S. Thesis, Northwestern Unversity, Evanston, IL, 1989, 65 pp.

Knight, J.H., Sensitivity of Time Domain Reflectometry Measurements to Lateral Variations in Soil Water Content, *Water Resour. Res.*, Vol. 28, 1992, pp. 2345-2352.

Knight, J.H., White, I., and Zegelin, S.J., Sampling Volume of TDR Probes for Water Content Monitoring, *Proceedings of the Symposium on Time Domain Reflectometry in Environmental, Infrastructure, and Mining Applications*, Evanston, Illinois, Sept 7-9, U.S. Bureau of Mines, Special Publication SP 19-94, 1994, NTIS PB95-105789, pp. 93-104.

Knowlton, R., Time Domain Reflectometry and Fiber Optic Probes for the Cone Penetrometer, Summary Rep., Workshop on Advancing Technologies for Cone Penetration Testing for Geotechnical and Geoenvironmental Site Characterization,

14-15 June, Eng. and Env. Sci. Div., U.S. Army Res. Office, Triangle Park, NC, Appendix F, 1994, 15 pp.

Knowlton, R., Onsurez, J., Bayliss, S., and Strong, W., Environmental Applications of TDR at Sandia National Labs, *Proceedings of the Symposium on Time Domain Reflectometry in Environmental, Infrastructure, and Mining Applications*, Evanston, Illinois, Sept 7-9, U.S. Bureau of Mines, Special Publication SP 19-94, 1994, NTIS PB95-105789, pp. 183-192.

Kotdawala, S.J., Hossain, M., and Gisi, A.J., Monitoring of Moisture Changes in Pavement Subgrades Using Time Domain Reflectometry (TDR), *Proceedings of the Symposium on Time Domain Reflectometry in Environmental, Infrastructure, and Mining Applications*, Evanston, Illinois, Sept 7-9, U.S. Bureau of Mines, Special Publication SP 19-94, 1994, NTIS PB95-105789, pp. 364-373.

Ledieu, J,, De Ridder, P., and Dautrebande, A., A Method for Measuring Soil Moisture by Time Domain Reflectometry, *J. Hydrology*, Vol. 88, 1986, pp. 319-328.

Lefter, J., Instrumentation for Measuring Scour at Bridge Piers and Abutments, *NCHRP Research Results Digest*, Transportation Research Board, No. 189, 1993, 8 pp.

Lin, Y., Gamble, W.L., and Hawkins, N.M., Report to ILLDOT for Testing of Bridge Piers, Poplar Street Bridge Approaches, Internal Report, Department of Civil Engineering, University of Illinois at Urbana-Champaign, 1994, 64 pp.

List, B.R., Stability Analysis Techniques Through the Monitoring of Syncrude's Open Pit Oil Sand Slopes, *Proceedings, 6th Int. Symp. on Landslides*, Christchurch, February, 1992, pp. 1299-1304.

Logan, J.W., Calibration of Time Domain Reflectometry Monitoring Cable in Granular Material, Report No. AFWL-NTE-TN-12-89, Civil Engineering Research Division, Air Force Weapons Laboratory, Kirtland AFB, March, 1989, 135 pp.

Logsdon, S., Calibrating TDR with Undisturbed Soil Cores for TDR Use in Monitoring Subsurface Lateral Water Flow, *Proceedings of the Symposium on Time Domain Reflectometry in Environmental, Infrastructure, and Mining Applications*, Evanston, Illinois, Sept 7-9, U.S. Bureau of Mines, Special Publication SP 19-94, 1994, NTIS PB95-105789, pp. 269-280.

Look, B.G., Reeves, I.N., and Williams, D.J., Field Experiences Using Time Domain Reflectometry for Monitoring Moisture Changes in Road Embankments and Pavements, *Proceedings of the Symposium on Time Domain Reflectometry in Environmental, Infrastructure, and Mining Applications*, Evanston, Illinois, Sept 7-9, U.S. Bureau of Mines, Special Publication SP 19-94, 1994a, NTIS PB95-105789, pp. 374-385.

Look, B.G., Reeves, I.N., and Williams, D. J., Application of Time Domain Reflectometry in the Design and Construction of Road Embankments, *Proceedings of the Symposium on Time Domain Reflectometry in Environmental, Infrastructure, and Mining Applications*, Evanston, Illinois, Sept 7-9, U.S. Bureau of Mines, Special Publication SP 19-94, 1994b, NTIS PB95-105789, pp. 410-421.

Lynch, L.J. and Webster, D.S., Effect of Thermal Treatment on the Interaction of Brown Coal and Water: a Nuclear Magnetic Resonance Study, *Fuel*, Vol. 61, 1982. p. 271.

Malicki, M.A., A Reflectometric (TDR) Meter of Moisture Content in Soils and Other Capillary-Porous Materials, *Zeszyty Problemowe Postepów Nauk Rolniczych* 388, 1990, pp. 107-114.

Malicki, M.A., Plagge, R., and Roth, C.H., Influence of Matrix on TDR Soil Moisture Readings and Its Elimination, *Proceedings of the Symposium on Time Domain Reflectometry in Environmental, Infrastructure, and Mining Applications*, Evanston, Illinois, Sept 7-9, U.S. Bureau of Mines, Special Publication SP 19-94, 1994, NTIS PB95-105789, pp. 105-114.

Marshall, C.E., *The Physical Chemistry and Mineralogy of Soils, Volume 1— Soil Materials*, John Wiley & Sons, New York, 1964.

McKenna, G., Advances in Slope Monitoring Instrumentation at Syncrude Canada Ltd.—a Follow-Up, *Geotechnical News*, Vol. 13, No. 1, March, 1995, pp. 42-44.

McKenna, G., Planning Geotechnical Tests and Investigations for Analysis and Presentation, *Geotechnical News*, Vol. 15, No. 1, March, 1997, pp. 40-44.

Measures, R.M., Lui, K., and Melle, S., Fiber-Optic Sensing System Critical Issues and Developments for Smart Structures, Active Materials and Adaptive Structures. *Proceedings of the ADPA/AIAA/ASME/SPIE Conference on Active Materials and Adaptive Structures*, G. J. Knowles, Ed., Nov., Alexandria, VA 1991.

Mehnert, B.B.,Van Roosendaal, D.J., Bauer, R.A., Barkley, D., and Gefell, E., Final report of Subsidence Investigations Over a High-Extraction Retreat Mine in Williamson County, Illinois, BuMines Cooperative Agreement CO267001, Illinois State Geological Survey, Sept.,1992, 91 pp.

Meltz, G., Morey, W.W., and Glen., W.H., Formation of Bragg Gratings in Optical Fiber by a Transverse Holographic Method, *Optics Lett.*, Vol. 14, 1989, p. 823.

Mickelson, A.R., Klevhus, O., and Eriksrud, M., Backscatter Readout from Serial Microbending Sensors, *Journal of Lightwave Technology*, Vol. LT-2, No. 5, Oct., 1984, pp. 700-709.

Mindess, S. and Young, J.F., *Concrete*, John Wiley & Sons, New York, 1981, p. 475.

Mine Monitoring Manual. Canadian Institute of Mining and Metallurgy, J. Franklin, Ed., Quebec, 1990.

Mitchell, J.K., *Fundamentals of Soil Behavior*, John Wiley & Sons, New York, 1976.

Mooijeer, H., *Microwave Techniques*, Macmillan, London, 1971.

Morey, W.W., Dunphy, J.R., and Meltz, G., Multiplexed Fiber Bragg Grating Sensors, *Proceedings, Distributed and Multiplexed Fiber Optic Sensors*, SPIE, Vol. 1586, 1991, p. 216.

Myers, B.K. and Marilley, J.M., Automated Monitoring at Tolt Dam, *Civil Engineering*, ASCE, Vol. 67, No. 3, March, 1997, pp. 44-46.

Neiber, J.L. and Baker, J.M., *In Situ* Measurement of Soil Water Content in the Presence of Freezing/Thawing Conditions, *State of the Art of Pavement Response Monitoring Systems for Roads and Airfields*, V. Janoo and R. Eaton, Eds., U. S. Army CRREL, Spec. Rep. 89-23, 1989.

Nieber, J.L., Baker, J.M., and Spaans, E.J.A., Evaluation of Soil Water Sensors in Frozen Soils, *Proc. Conf. on Road and Pavement Response Monitoring Systems*, ASCE, New York, 1991, pp. 168-181.

Nyhan, J.W., Schofield, T.G., and Martin, C.E., Use of Time Domain Reflectometry in Hydrologic Studies of Multilayered Landfill Covers for Closure of Waste Landfills at Los Alamos, New Mexico, *Proceedings of the Symposium on Time Domain Reflectometry in Environmental, Infrastructure, and Mining Applications*, Evanston, Illinois, Sept 7-9, U.S. Bureau of Mines, Special Publication SP 19-94, 1994, NTIS PB95-105789, pp.193-206.

O'Connor, K.M., Monitoring of Rock Response to Solution Mining, State Min. and Miner. Res. Res. Inst., Grant 9C-BD-03-9-0000, New Mexico Tech, Socorro, June, 1989, 26 pp.

O'Connor, K.M., Development of a System for Highwall Monitoring Using Time Domain Reflectometry, U.S. Bureau of Mines Sum. Rep., prepared for Syncrude Research Ltd., Edmonton, Aug., 1991, 75 pp.

O'Connor, K.M., Overburden Response to Longwall Mining, Emery County, Utah, Compiled TDR waveforms, U.S. Bureau of Mines, Minneapolis, 1992, 370 Kbyte file.

O'Connor, K.M., Overburden Response to Longwall Mining, Oakland, Maryland, Compiled TDR waveforms, U.S. Bureau of Mines, Minneapolis, 1994, 81 Kbyte file.

O'Connor, K.M., Remote Detection of Strata Movements over Abandoned Coal Mines, Abandoned Mine Lands Research, Final Report, USBM, Minneapolis, MN, November,1995,

93 pp. and 2 disks.

O'Connor, K.M., Demonstration of Water Level Detection at an Existing Dam Using TDR, Internal Report, GeoTDR, Inc., Apple Valley, MN, 1996, 25 pp.

O'Connor, K.M., Dowding, C.H., and Jones, C.C., Editors. *Proceedings of the Symposium on Time Domain Reflectometry in Environmental, Infrastructure, and Mining Applications*, Evanston, Illinois, Sept 7-9, U.S. Bureau of Mines, Special Publication SP 19-94, 1994, NTIS PB95-105789, 665 pp.

O'Connor, K.M. and Norland, M.R., Monitoring Subsidence Mechanisms Using Time Domain Reflectometry, *Proceedings, Joseph F. Poland Symposium on Land Subsidence*, Assoc. of Engineering Geologists, Sacramento, October, 1995, pp. 427-434.

O'Connor, K.M., O'Rourke, J.E., and Carr, J., Influence of Rock Discontinuities on Coal Mine Subsidence, *BuMines OFR 30-84*,1983, NTIS PB 84-166370, 129 pp.

O'Connor, K.M., Peterson, D.E., and Lord, E.R., Development of a Highwall Monitoring System using Time Domain Reflectometry, *Proc., 35th U.S. Sym. Rock Mech.*, Reno, Nevada, June, 1995, pp. 79-84.

O'Connor, K.M. and Zimmerly, T., Application of Time Domain Reflectometry to Ground Control, *Paper in Proceedings of the 10th International Conference on Ground Control in Mining* (Morgantown, June, 1991), WV Univ., 1991, pp. 115-121.

O'Rourke, J.E., Rey, P.H., and O'Connor, K.M., Characterization of Subsidence over Longwall Panels, *BuMines OFR 3-84*, 1982, NTIS PB 84-142918, 180 pp.

Opto-Electronics (Oakville, Ontario), *Picosecond Fiberoptic Instruments Catalog*, 1994.

Panek, L.A. and Tesch, W.J., Monitoring Ground Movements Near Caving Stopes-Methods and Measurements, *BuMines RI 8585*,1981, 109 pp.

Paquet, J.M., Caron, J., and Banton, O., *In Situ* Determination of the Water Desorption Characteristics of Peat Substances, *Can. J. Soil Sci.*, Vol. 73, 1993, pp. 329-339.

Park, R. and Paulay, T., *Reinforced Concrete Structures*, John Wiley & Sons, New York, 1975.

Patterson, D.E. and Smith, M.W., The Use of Time Domain Reflectometry for the Measurement of Unfrozen Water Content in Frozen Soils, *Cold Regions Science and Technology*, Vol. 3, 1980, pp. 205-210.

Patterson, D.E. and Smith, M.W., The Measurement of Unfrozen Water Content by Time Domain Reflectometry: Results from Laboratory Tests, *Canadian Geotechnical Journal*, Vol. 18, 1981, pp. 131-144.

Peck, R.B., Hanson, W.E., and Thornburn, T.H., *Foundation Engineering*, 2nd ed., John Wiley & Sons, Inc, New York, 1974, 271.

Pepin, S., Plamondon, A.P., and Stein, J., Peat Water Measurement Using Time Domain Reflectometry, *Can. J. For. Res.*, Vol. 22, 1992, pp. 534-540.

PermAlert ESP (Niles, Illinois), *Leak Detection/Location Systems*, 1995. 10 pp.

Petersen, L.W., Sampling Volume of TDR Probes Used for Water Content: Practical Investigation, *Paper in the Proceeedings of the Symposium: Time-Domain Reflectometry Applications in Soil Science* (Foulum, Denmark, Sept 16, 1994) SP Report No. 11, June, Vol. 3, 1995, pp. 57-62.

Peterson, D., Personal communication, Syncrude Canada Ltd., Edmonton, Alberta, April, 1993.

Pierce, C.E. and Dowding, C.H., Long-Term Monitoring of Bridge Pier Integrity with Time Domain Reflectometry Cables, *Proc. of Conference and Exposition of Sensors and Systems*, Sensors Magazine, 1995. pp. 399-406.

Pierce, C.E., Bilaine, C., Huang, F.-C., and Dowding, C.H., Effects of Multiple Crimps and Cable Length on Reflection Signatures from Long Cables, *Proceedings of the Symposium on Time Domain Reflectometry in Environmental, Infrastructure, and Mining Applications*, Evanston, Illinois, Sept 7-9, U.S. Bureau of Mines, Special Publication SP 19-94, 1994, NTIS PB95-105789, pp. 540-554.

Pulse, R.R., Results of the Subsidence Study at Old Ben No. 21; (Oct 68 - July 70) Settlement Probe, BPC & TDR Monitoring Methods, BuMines, Denver, July,1970, 4 pp.

Rada, G.R., Lopez, A., and Elkins, G.E., Monitoring of Subsurface Moisture in Pavements using Time Domain Reflectometry, *Proceedings of the Symposium on Time Domain Reflectometry in Environmental, Infrastructure, and Mining Applications*, Evanston, Illinois, Sept 7-9, U.S. Bureau of Mines, Special Publication SP 19-94, 1994, NTIS PB95-105789, pp. 422-433.

Ramo, S. and Whinnery, J.R., *Fields and Waves in Modern Radio*, John Wiley & Sons, New York, 1959.

Ramo, S., Whinnery, J.R., and VanDuzer, T., *Fields and Waves in Communication Electronics*, John Wiley & Sons, New York, 1965.

Reeves, T.L. and Elgezawi, S.M., Time Domain Reflectometry for Measuring Volumetric Water Content in Processed Oil Shale Waste, *Water Resources Research*, Vol. 28, No. 3, March, 1992, pp. 769-776.

Ross, G.F., Apparatus and Method for Measuring the Level of a Contained Liquid, U.S. Patent No. 3,832,900, 1974.

Ross, G.F., Apparatus and Method for Sensing a Liquid with a Single Wire Transmission Line, U.S. Patent No. 3,995,212, 1976.

Roth, C.H., Malicki, M.A., and Plagge, R., Empirical Evaluation of the Relationship Between Soil Dielectric Constant and Volumetric Water Content as a Basis for Calibrating Soil Moisture Measurements, *J. Soil Sci.*, Vol. 43, 1992, pp.1-13.

Roth, K.R., Schulin, R., Flühler, H., and Attinger, W., Calibration of Time Domain Reflectometry for Water Content Measurement Using a Composite Dielectric Approach, *Water Resour. Res.*, Vol. 26, 1990, pp. 2267-2273.

Schofield, T.G., Langhorst, G.J., Trujillo, G., Bostick, K.V., and Hansen, W.R., Comparison of Neutron Probe and Time Domain Reflectometry Techniques of Soil Moisture Analysis, *Proceedings of the Symposium on Time Domain Reflectometry in Environmental, Infrastructure, and*

Mining Applications, Evanston, Illinois, Sept 7-9, U.S. Bureau of Mines, Special Publication SP 19-94, 1994, NTIS PB95-105789, pp. 130-142.

Selker, J.S., Graff, L., and Steenhuis, T., Noninvasive Time Domain Reflectometry Moisture Measurement Probe, *Soil Sci. Soc. Am. J.,* Vol. 57, 1993, pp. 934-936.

Sevick, J., *Transmission Line Transformers,* American Radio Relay League, Newington, CT, 1990.

Siekmeier, J.A., O'Connor, K.M., and Powell, L.R., Rock Mass Classification Applied to Subsidence Over High Extraction Coal Mines, *Paper in Proceedings of the Third Workshop on Subsidence Due to Underground Mining* (Morgantown, June, 1992). WV Univ., 1992, pp. 317-325.

Siekmeier, J.A. and O'Connor, K.M., Description of Laminated Overburden Using Rock Mass Classification, *Proceedings, Fifth Conf. on Ground Control for Midwestern Coal Mines,* June, 1994, pp. 1-13.

Sivhola, A. and Lindell, I.V., Polarizability and Effective Permittivity of Layered and Continuously Inhomogeneous Dielectric Spheres, *J. Electromagnetic Waves and Applications,* Vol. 2, No. 8, 1988, pp. 741-756.

Skaling, W., TRASE: A Product History, *Advances in Measurement of Soil Physical Properties: Bringing Theory into Practice,* G. C. Topp et al., Eds., Soil Sci. Soc. Am., Madison, WI, SSSA Sp. Pub. 30, 1992, pp. 169-185.

Smith, C., OTDR Technology and Optical Sensors, *Proceedings of the Symposium on Time Domain Reflectometry in Environmental, Infrastructure, and Mining Applications,* Evanston, Illinois, Sept 7-9, U.S. Bureau of Mines, Special Publication SP 19-94, 1994, NTIS PB95-105789, pp. 14-22.

Society of the Plastics Industry, *Plastics Engineering Handbook,* 3d ed., Reinhold Publishing, New York, 1960.

Soilmoisture Equipment Co. (Santa Barbara, CA), *TRASE. Instruments for the Measurement of Moisture and Dielectric Properties Using Time Domain Reflectometry,* 1995.

Spaans, E.J.A. and Baker, J.M., Simple Baluns in Parallel Probes for Time Domain Reflectometry, *Soil Sci. Soc. Am.*, Vol. 57, 1993, pp. 668-673.

Spergel, J., Coaxial Cable and Connector Systems, *Handbook of Wiring, Cabling, and Interconnecting for Electronics*, C. A. Harper, Ed., McGraw-Hill, New York, 1972, Chapter 4.

Stein, J. and Kane, D.L., Monitoring the Unfrozen Water Content of Soil and Snow Using Time Domain Reflectometry, *Water Resources Research*, Vol. 19, No. 6, Dec., 1983, pp. 1573-1584.

Sterling, D.J., *Technician's Guide to Fiber Optics*, Delmar Publishers, Inc., 1987. ISBN 0-8273-2614-9.

Su, M.B., Quantification of Cable Deformation with Time Domain Reflectometry, Ph.D. Dissertation, Northwestern Univ., Evanston, IL, 1987, 112 pp.

Su, W.H. and Hasenfus, G.J., Field Measurements of Overburden and Chain Pillar Response to Longwall Mining, *Paper in Proceedings of the 6th International Conference on Ground Control in Mining* (Morgantown, 1987), WV Univ., 1987, pp. 296-311.

Taflove, A., *Computational Electrodynamics: The Finite-Difference Time-Domain Method*, Artech House, Boston, 1995. 599 pp.

Tarnoff, S., Surcharge to Refill Tank Clean-up Fund, *Crain's Small Business*, Chicago, IL, June, 1996, p. 4.

Tektronix, Inc. (Redmond, OR), *1502B Metallic Time Domain Reflectometer Operator Manual*, 1989a.

Tektronix, Inc. (Redmond, OR), *SP232 Serial Extended Function Module Operator/Service Manual*, Jan., 1989b.

Tektronix, Inc. (Redmond, OR), *Optical Reflection Testing Using the OIG502.* Applications Note, 1992.

Tektronix, Inc. (Redmond, OR), *TFP2 FiberMaster Operator Manual*, 1993.

Terzaghi, K. and Peck, R.B., *Soil Mechanics in Engineering Practice*, 2nd ed., John Wiley, New York, 1967.

Topp, G.C. and Davis, J.L., Time Domain Reflectometry (TDR) and its Application to Irrigation Scheduling, *Advances in Irrigation*, Vol. 3, Academic Press, 1985, pp. 107-127.

Topp, G.C., Davis, J.L., and Annan, A.P., Electromagnetic Determination of Soil Water Content: Measurement in Coaxial Transmission Lines, *Water Resources Research*, Vol. 16, No. 3, June, 1980, pp. 574-582.

Topp, G.C., Davis, J.L., and Annan, A.P., Electromagnetic Determination of Soil Water Content Using TDR: I. Applications to Wetting Fronts and Steep Gradients, *Soil Sci. Soc. Am. J.*, Vol. 46, 1982, pp. 672-684.

Topp, G.C., Davis, J.L., Bailey, W.G., and Zebchuk, W.D., The Measurement of Soil Water Content Using a Portable TDR Hand Probe, *Can. J. Soil Sci.*, Vol. 64,1984, pp. 313-321.

Topp, G.C., Davis, J.L., and Chinnick, J.H., Using TDR Water Content Measurements for Infiltration Studies, *Advances in Infiltration*, Pub. 11-83, Am. Soc. Ag. Eng., 1983, pp. 231-240.

Topp, G.C., Yanuka, M., Zebchuk, W.D., and Zegelin, S., Determination of Electrical Conductivity Using Time Domain Reflectometry: Soil and Water Experiments in Coaxial Lines, *Water Resources Research*, Vol. 24, 1988, pp. 945-952.

Topp, G.C., Zegelin, S.J., and White, I., Monitoring Soil Water Content Using TDR: An Overview of Progress, *Proceedings of the Symposium on Time Domain Reflectometry in Environmental, Infrastructure, and Mining Applications*, Evanston, Illinois, Sept 7-9, U.S. Bureau of Mines, Special Publication SP 19-94, 1994, NTIS PB95-105789, pp. 67-80.

Torres, R., Dietrich, W.E., Anderson, S.P., Montgomery, D.R., and Loague, K., Experimental Observations of Unsaturated Flow and Soil-Water Distribution at the Catchment Scale, *EOS Supplement*, 27 October, Am. Geophy. Union, Fall Meeting, 1992, p. 209.

Triplett, T.L., Bennett, R.D., and Murphy, E.W., Demonstration of Bond Beam Technology for Construction of Residential Foundations Resistant to Subsidence Damage, Sum. Rep., Joint Project of the U. S. Bureau of Mines and Illinois Mine Subsidence Insurance Fund, Nov., 1995.

Vadose Zone Equipment Co. (Amarillo, TX), *Products for Time Domain Reflectometry*, 1994.

van der Keur, P., Measuring Unfrozen Water Content in Partially Frozen Soil Using TDR Technique, *Paper in the Proceedings of the Symposium: Time-Domain Reflectometry Applications in Soil Science*, (Foulum, Denmark, Sept 16, 1994) SP Report No. 11, June, Vol. 3, 1995, pp. 133-137.

Van Krevelin, D.W., *Coal: Typology, Chemistry, Physics and Constitution*, Elsevier, New York, 1981.

van Loon, W.K.P., Smulders, P.E.A., van den Berg, C., and van Haneghem, I.A., Time-domain Reflectometry in Carbohydrate Solutions, *Journal of Food Engineering*, Vol. 26, 1995, pp. 319-328.

Van Roosendaal, D.J., Brutcher, D.F., Mehnert, B.B., Kelleher, J.T., and Bauer, R.A., Overburden Deformation and Hydrologic Changes Due to Longwall Mine Subsidence in Illinois, *Paper in Proceedings of the Third Conference on Ground Control Problems in the Illinois Coal Basin* (Carbondale, June, 1990), Southern IL. Univ., 1990, pp. 73-82.

Van Roosendaal, D.J., Mehnert, B.B., and Bauer, R.A., Three-Dimensional Ground Movements During Dynamic Subsidence of a Longwall Mine in Illinois, *Paper in Proceedings of the Third Workshop on Surface Subsidence Due to Underground Mining* (Morgantown, June 1-4, 1992), WV Univ, 1992, pp. 290-298.

van Schelt, W., de Jong Hänninen, M.K., van der Aa, J.P.C.M., and van Loon, W.K.P., Field and Laboratory Experiments with Time Domain Reflectometry as a Moisture Monitoring System in Road Structures, *Proceedings of the Symposium on Time Domain Reflectometry in Environmental, Infrastructure, and Mining Applications*, Evanston, Illinois, Sept 7-9, U.S. Bureau of Mines, Special Publication SP 19-94, 1994, NTIS PB95-105789, pp. 386-397.

Van Wesenbeek, I.J. and Kachanoski, R.G., Spatial and Temporal Distribution of Soil Water in the Tilled Layer Under a Corn Crop, *Soil Sci. Soc. Am. J.*, Vol. 52, 1988, pp. 363-368.

Verstricht, J., Neerdael, B., Meynendonckx, P., and Volckaert, G., Clay Moisture Measurements in Radioactive Waste Disposal Research, *Proceedings of the Symposium on Time Domain Reflectometry in Environmental, Infrastructure, and Mining Applications*, Evanston, Illinois, Sept 7-9, U.S. Bureau of Mines, Special Publication SP 19-94, 1994, NTIS PB95-105789, pp. 337-348.

Wade, L.V. and Conroy, P.J., Rock Mechanics Study of a Longwall Panel. *Mining Engineering*, Dec., 1980, pp. 1728-1734.

Wagner, K.W., The After Effect in Dielectrics, *Arch. Elektrotech Berlin*, Vol. 2, 1914, pp. 371-387.

Wanser, K.H., Voss, K.F., and Kersey, A.D., Multimode Fiber Bragg Gratings for Real Time Structural Monitoring Using Optical Time Domain Reflectometry, *Proceedings of the Symposium on Time Domain Reflectometry in Environmental, Infrastructure, and Mining Applications*, Evanston, Illinois, Sept 7-9, U.S. Bureau of Mines, Special Publication SP 19-94, 1994, NTIS PB95-105789, pp. 472-483.

Weaver, K., *Dam Grouting*, ASCE, New York, 1991.

Whalley, W.R, Considerations on the Use of Time Domain Reflectometry (TDR) for Measuring Soil Water Content, *J. Soil Sci.*, Vol. 44, 1993, pp. 1-9.

White, I., Zegelin, S.J., Topp, G.C., and Fish, A., Effect of Bulk Electrical Conductivity on TDR Measurement of Water Content in Porous Media, *Proceedings of the Symposium on Time Domain Reflectometry in Environmental, Infrastructure, and Mining Applications*, Evanston, Illinois, Sept 7-9, U.S. Bureau of Mines, Special Publication SP 19-94, 1994, NTIS PB95-105789, pp. 294-308.

Wobschall, D., A Theory of the Complex Dielectric Permittivity of Soil Containing Water: The Semi-Disperse Model, *IEEE Trans. Geosci. Electron.*, Vol. 15, 1977, pp. 49-58.

Xanthakos, P.P., Abramson, L.W., and Bruce, D.A., *Ground Control and Improvement*, John Wiley & Sons, New York, 1994.

Yanuka, M., Topp, G.C., Zegelin, S., and Zebchuk, W.D., Multiple Reflections and Attenuation of Time Domain Reflectometry Pulses: Theoretical Considerations for Applications to Soil and Water, *Water Resources Research*, Vol. 24, 1988, pp. 934-944.

Yong, R.N. and Hoppe, E.J., Application of Electric Polarization to Contaminant Detection in Soils, *Canadian Geotechnical Journal*, Vol. 26, No. 4, Nov., 1989, pp. 536-550.

Zegelin, S. J. and White, I., Calibration of TDR for Applications in Mining, Grains, and Fruit Storage and Handling, *Proceedings of the Symposium on Time Domain Reflectometry in Environmental, Infrastructure, and Mining Applications*, Evanston, Illinois, Sept 7-9, U.S. Bureau of Mines, Special Publication SP 19-94, 1994, NTIS PB95-105789, pp. 115-129.

Zegelin, S.J., White, I., and Jenkins, D.R., Improved Field Probes for Soil Water Content and Electrical Conductivity Measurement Using Time Domain Reflectometry, *Water Resources Research*, Vol. 25, No. 11, Nov., 1989, pp 2367-2376.

Zegelin, S.J., White, I., and Russel, G.F., A Critique of the Time Domain Reflectometry Technique for Determining Soil Water Content, *Advances in the Measurement of Soil Physical Properties: Bringing Theory into Practice*, G. C. Topp et al., Eds., Soil Sci. Soc. Am., Madison, WI, SSSA Sp. Pub. 30, 1992, pp. 187-208.

APPENDIX C
LIST OF SYMBOLS

GREEK SYMBOLS

α attenuation coefficient (nepers/m or decibels)

α angle from vertical axis of tension link in TDR calibration apparatus (radians)

α_{eff} effective attenuation coefficient (nepers/m)

β geometric factor in dielectric mixing formula

γ propagation constant (sec/m)

γ_{bulk} bulk unit weight (N/m^3)

γ_d unit weight of dry solids (N/m^3)

γ_w unit weight of water (981 N/m^3 @ 20°C)

Δ smallest detectable change in distance to air–water interface (m)

ϵ dielectric permittivity ($\epsilon = K\epsilon_0$) (farad/m)

ϵ' real part of dielectric permittivity (farad/m)

ϵ'' imaginary part of dielectric permittivity (farad/m)

ϵ_0 dielectric permittivity of free space ($10^{-9}/36\pi = 8.85 \times 10^{-12}$ farad/m)

ϵ_a dielectric permittivity of air (farad/m)

ϵ_w dielectric permittivity of water (farad/m)

η void ratio (m^3/m^3)

θ angle from horizontal axis of tensile force in TDR calibration apparatus

θ_{bw} volumetric content of bound water (m^3/m^3)

θ_{fw} volumetric content of free water (m^3/m^3)

θ_m gravimetric water content (N/N or kg/kg)

θ_v volumetric water content (m^3/m^3)

θ_{uf} volumetric content of unfrozen water (m^3/m^3)

λ_B wavelength for which Bragg resonance is satisfied (m)

Λ spatial pitch of Bragg grating written into an optical fiber (m)

μm micron (10^{-6} m)

μ magnetic permeability (henry/m)

μ_0 magnetic permeability of free space
         ($4\pi \times 10^{-7} = 1.26 \times 10^{-6}$ henry/m)
π pi (= 3.14)
ρ reflection coefficient, rho
ρ_b bulk density of moist soil (g/cm^3)
ρ_d density of dry soil (g/cm^3)
ρ_w density of water (1.0 g/cm^3 @ 20°C)
σ conductivity (Siemens)
σ_{dc} static or dc conductivity (siemens)
σ_x, σ_y normal stress (N/m^2)
σ_{xy} shear stress (N/m^2)
τ exponential time constant (sec)
τ_c capacitive time constant (sec)
τ_L inductive time constant (sec)
ϕ_{peak} peak friction angle (degrees)
ϕ_{res} residual friction angle (degrees)
ψ soil water potential (N/m^2)
ω angular frequency (radians/sec)

ENGLISH SYMBOLS

a   radius of outer conductor of coaxial cable (m)
a   proportionality constant between water potential and water content
a_1, a_2 constants for logarithmic regression of displacement versus TDR reflection magnitude
b   radius of inner conductor of coaxial cable (m)
b   geometric constant relating water content to water potential
B_w wheel base length of inclinometer (m)
c   speed of an electromagnetic wave in free space
         (3 x 10^8 m/s) [= 1 / $(\mu_0\epsilon_0)^{1/2}$]
c   cohesion component of shear strength (N/m^2)
C   capacitance per unit length (farads/m)
C_0 initial capacitance of coaxial cable (farads/m)
C_1 capacitance of deformed coaxial cable (farads/m)
C_s shunt capacitance (farads)
C_t terminal capacitance (farads)

d   diameter of TDR probe rods (m)
D   borehole diameter (m)
D_f embedment depth (m)
D_{pt} height of water in standpipe (MPa or N/m^2)
D_r relative density of soil (%)
D_s depth to porous stone or screen in a standpipe (m)
E   shear stiffness of grout (N/m^2)
f   frequency (cycles/sec or Hertz)
f( )   "function of"
G   conductance per unit length (siemens/m)
G_s specific gravity
Hz   Hertz = cycles/sec
i   sample location (m)
I   current (amperes)
I_o moment of inertia of grout–cable cross-sectional area around neutral axis (m^4)
j   square root of -1
k_s coefficient of subgrade reaction ($N/m/m^2$)
K   dielectric constant ($\epsilon = K\epsilon_0$)
K_a apparent dielectric constant
K_{air} dielectric constant of air
K_{bw} dielectric constant of bound water
K_{fw} dielectric constant of free water
K_{ice} dielectric constant of ice
K_s dielectric constant of solids
K_{water} dielectric constant of water
l   liter
l_a apparent length of TDR probe (m)
l_p actual length of TDR probe rods (m)
l_s actual submerged cable length (m)
l_w pulse travel length to air–water interface (m)
L   inductance per unit length (henries/m)
L_o soil-structure interaction transfer distance (m)
L_s series inductance (henries)
L_t terminal inductance (henries)
n   effective index of refraction of glass fiber
n_{air} index of refraction of air with respect to air (at standard conditions and wavelength)

n_{glass} index of refraction of glass with respect to air (at standard conditions and wavelength)

N_γ bearing area shape factor of general bearing capacity equation

N_c shear strength factor of general bearing capacity equation

N_q embedment depth factor of general bearing capacity equation

N_Q number of streamlines

N_V number of flow lines

p passive resistance developed by soil acting on grouted cable (N/m^2)

P compressive load applied to TDR calibration apparatus (N)

q bearing capacity of soil (N/m^2)

q_u unconfined, undrained compressive strength (Pa or N/m^2)

R resistance per unit length (ohms/m)

R_t termination resistance (ohms)

s spacing of TDR probe rods (m)

S sensitivity of TDR reflection magnitude to cable deformation (mrho/mm)

t time (sec)

t_r rise time (sec)

T temperature (°C)

T_b, T_c tensile forces in TDR calibration apparatus (N)

u displacement transverse to axis of coaxial cable (mm)

u_0 displacement required before TDR reflection is detected

u_w pore water pressure (kPa)

v normal or extension displacement (mm)

V voltage (volts)

V_o voltage output from TDR pulser (volts)

V_1 transmitted voltage (volts)

V_2 reflected voltage after one return trip (volts)

V_{air} propagation velocity of wire or cable in air (% of c)

V_{bubble} propagation velocity of wire or cable covered with residual water drops (% of c)

V_f voltage after all reflections have occurred (volts)

V_i incident voltage (volts)

V_p propagation velocity (percentage of the speed of light, c)

V_{pk} peak voltage (volts) or TDR reflection magnitude (mrho)

V_{pk0}　　y-axis intercept of linear regression between TDR reflection magnitude and displacement (mρ or mrho)

$V_{Ptester}$　propagation velocity setting on cable tester

V_r　　　reflected voltage (volts)

V_s　　　shear stress within grout-cable column (N/m^2)

V_{tot}　　total volume (m^3)

V_W　　　volume of water (m^3)

w　　　　gravimetric water content (N/N or Kg/Kg)

W_s　　　weight of dry solids (N)

W_{tot}　　total weight of sample (N)

W_W　　　weight of water (N)

X_1　　　uncorrected TDR transmission distance (m)

X_2　　　corrected transmission distance (m)

X_M　　　physically measured distance to the air–water interface (m)

X_D　　　shortest distance, corresponding to the most recent maximum height of water (m)

Y　　　　admittance (siemens)

Z　　　　impedance (ohms)

Z_0　　　characteristic impedance of transmission line (ohms)

Z_1　　　impedance of deformed coaxial cable (ohms)

Z_t　　　impedance of termination (ohms)

Z_u　　　impedance of a filled coaxial sample holder (ohms)

UNIT OF MEASURE ABBREVIATIONS

amp	coul/sec
bar	10^5 N/m^2
bps	bits per second
cm	centimeter
cm/day	centimeter per day
cm/hr	centimeter per hour
cm/s	centimeter per second
cm^3/cm^3	cubic centimeter per cubic centimeter
cm^3/h	cubic centimeter per hour
coul	amp-sec
db	decibel
dB/km	decibel per kilometer

dS/m	decasiemens/meter
°C	degree Celsius
°K	degree Kelvin
farad	coul/volt
ft	foot
ft/div	foot per division
g/cm^3	gram per cubic centimeter
g/g	gram per gram
g/L	gram per liter
g/m^2	gram per square meter
GHz	gigahertz (10^9 cycles/sec)
ha	hectare
henry	volt-sec/amp
hr	hour
in.	inch
in./yr	inch per year
joule	N-m
kg/m^2	kilogram per square meter
kg/m^3	kilogram per cubic meter
km	kilometer (10^3 m)
kN	kilonewton
kPa	kilopascal
l/hr	liter per hour
m	meter
m/s	meter per second
m^{-1}	per meter
m^3/m^3	cubic meter per cubic meter
Mg/m^3	megagram per cubic meter
MHz	megahertz (10^6 cycles/sec)
mm	millimeter (10^{-3} m)
mm/day	millimeter per day
mm/yr	millimeter per year
Mohm	megaohm
MPa	megapascal (10^6 Pa)
mρ	millirho
mrho	millirho
mS/cm	millisiemens per centimeter
nm	nanometer (10^{-9} m)

ns	nanosecond (10^{-9} sec)
Ω	ohm (volt/amp)
Pa	pascal (N/m^2)
pcf	pound per cubic foot
ps	picosecond (10^{-12} sec)
psi	pound per square inch
ρ	rho
S/m	siemens per meter (10 millimho/cm)
V	volt (joule/coul)
W	watt (joule/sec)

ACRONYMS USED

AASHTO	American Association of State Highway and Transportation Officials
AC	alternating current
ACP	asphalt concrete pavement
APD	avalanche photo diode
ASCE	American Society of Civil Engineers
ASTM	American Society for Testing and Materials
BNC	coaxial connector
CALTRANS	California State Department of Transportation
CANMET-MRL	Canadian Centre for Mineral and Energy Technology–Mining Research Laboratory
CBR	California Bearing Ratio
CL	low plasticity clay
COE	U.S. Army Corps of Engineers
CPAR	Construction Productivity Advancement Research
CRREL	U.S. Army Cold Regions Research and Engineering Laboratory
CSIRO	Commonwealth Scientific and Industrial Research Organization (Australia)
CST	cable shear tester
d.c.	direct current
DC	direct current
EPA	U.S. Environmental Protection Agency
FAO	family and order of soil classification

FBG	optical fiber Bragg grating
FD	frequency domain
FFT	Fast Fourier Transform
FHWA	Federal Highway Administration
FO	fiber optical
FXA	Foamflex coaxial cable (Cablewave Systems, New Haven, CT)
HF	High Frequency (3 to 30 MHz)
IEEE	Institute of Electrical and Electronic Engineeers
IWP	inner wheel path
laser	Light Amplification by Stimulated Emission of Radiation
LC	inductance–capacitance
LED	light emitting diode
lidar	Light RADar
LVDT	linear variable differential transformer
MCU	multiplexed control unit (Geomation)
MDD	maximum dry density
MPBX	multiple point borehole extensometer
MTDR	Metallic Time Domain Reflectometry
N-type	coaxial connector
NCHRP	National Cooperative Highway Research Program
NDE	nondestructive evaluation
NSF	National Science Foundation
NSW	New South Wales
NUTSA	Northwestern University TDR Signature Analysis Program
OD	outside diameter
OEM	original equipment manufacturer
OFDR	Optical Frequency Domain Reflectometry
OMC	optimum moisture content
OTDR	Optical Time Domain Reflectometry
OWP	outer wheel path
PC	personal computer
PIN	positive-intrinsic-negative
PNET	Peaceful Nuclear Explosion Treaty
PROM	Programmable Read Only Memory
PSPL	Picosecond Pulse Labs

PVC	polyvinyl chloride
PYELAB	TDR system developed by CSIRO
radar	RAdio Detecting And Ranging
RF	Radio Frequency (0.3 to 30000 MHz; AM 0.55 to 1.6 MHz; FM 88 to 108 MHz)
RG/U	RG is military designation for coaxial cable, U stands for general utility
RJ-11	IEEE standard for telecommunication
RS232	IEEE standard for serial communication
SHRP	Strategic Highway Research Program
SINCO	Slope Indicator Company
SLC	grout mix used by Syncrude Canada, Ltd.
SMA	coaxial connector
SMP	Seasonal Monitoring Program
SPDV	Site Preliminary Design Validation
SPHOST	Host Application Program for Serial Communication with SP232 serial module (Tektronix)
SPICE	simulation and analysis program developed by Intusoft, San Pedro, CA
SPIE	The International Society for Optical Engineering
SPT	Standard Penetration Test
SSSA	Soil Science Society of America
TDR	Time Domain Reflectometry
TEM	transverse electric and magnetic mode
TRASE	Time Reflectometry Analysis of Signal Energy (TDR system developed by Soil Moisture Equipment, Inc.)
TRIME	TDR system developed by IMKO
TV	television (VHF 54 to 72 MHz; UHF 470 to 806 MHz)
UHF	UltraHigh Frequency (300 to 3000 MHz)
USDA	U.S. Department of Agriculture
USBM	U.S. Bureau of Mines
USGS	U.S. Geological Survey
UV	ultraviolet (10^{10} to 10^{11} MHz)
VHF	Very High Frequency (30 to 300 MHz)
WIPP	Waste Isolation Pilot Plant

APPENDIX D
AUTHOR INDEX

APPENDIX E
SUBJECT INDEX

APPENDIX F
VENDORS

CABLE

Anixter Bros., Inc., 4711 Golf Road, Skokie, IL 60076-9910
(847) 677-2600, (847) 677-9480 FAX
http://www.anixter.com

Belden Wire and Cable, P.O. Box 1980, Richmond, IN 47375
(317) 983-5200, (800) BELDEN-1, (317) 983-5295 FAX
http://www.belden.com

Cablewave Systems Division, Radio Frequency Systems, Inc.,
60 Dodge Avenue, North Haven, CT 06473
(203) 239-3311, (203) 234-7718 FAX
http://www.cablewave.com

CommScope, Inc. of North Carolina, P.O. Box 1729, Hickory,
NC 28603-1729
(800) 982-1708, (704) 328-3400 FAX
http://www.commscope.com

Hergalite: Anthem Marketing Co., 2525-7 Pioneer Ave., Vista,
CA 92083
(760) 734-6996, (760) 734-6829 FAX

SpecTran Corp., 50 Hall Rd, Sturbridge, MA 01566
(508) 347-2261, (508) 347-1211 FAX
http://www.spectran.com

BALUNS

Communication Coil, Inc., 9601 Soreng Avenue, Schiller Park,
IL 60176
(847) 671-1333, (847) 671-9191 FAX

Midwest Special Services, 900 Ocean St., St. Paul, MN 55106
(612) 778-1000, (612) 772-5352 FAX

GROUT ADDITIVES

Intrusion Aid Type R, Specrete-IP Inc., 10703 Quebec Avenue,
Cleveland, OH 44106
(216) 721-2050, (216) 421-0032 FAX

EQUIPMENT FOR TDR MEASUREMENTS

Bicotest Limited, Delamare Road, Cheshunt,
Hertfordshire EN8 9TG, UNITED KINGDOM
Contact:Wendy Seegar
(44) 1992 629011, (44) 1992 636170 FAX
email: info@bicotest.demon.co.uk
http://www.bicotest.co.uk

Biddle Instruments, AVO International, 510 Township Line Road
Blue Bell, PA 19422-2795
Contact: Tony Kratowicz
(800) 723-2861, (215) 283-2215, (215) 643-2670 FAX
email: tony.kratowicz@avointl.com
http://www.avointl.com

Campbell Scientific, Inc., 815 W. 1880 N.,
Logan, UT 84321-1784
Contact: Jim Bilskie
(801) 753-2342, (801) 752-3268 FAX
http://www.campbellsci.com

Campbell Scientific Australia P/L, PO Box 444, Thuringowa
Central, Kirawn, Queensland, 4817 AUSTRALIA
Contact: Jason Beringer or Steve Bailey
(61) 77 254100, (61) 77 254155 FAX
email: csa@ultra.net.au
http://www.campbellsci.com

Campbell Scientific Ltd., 80 Hathern Road, Shepshed
Leicestershire LE12 9RP UNITED KINGDOM
Contact: Andrew Sandford
(44) 0 1509 601141, (44) 0 1509 601091 FAX
email: andrew@campbellsci.co.uk
http://www.csluk.demon.co.uk

CM Technologies Corporation, 1026 Fourth Avenue,
Coraopolis, PA 15108
Contact: Greg Allan
(412) 262-0734, (412) 262-2250 FAX
email: ecadusa@aol.com

Dynamax, Inc., 10808 Fallstone, Suite 350, Houston, TX 77099
Contact: Michael McClung
(800) 727-3570 (domestic), (281) 564-5100,
(281) 564-5200 FAX
email: dynamax@mail.pernet.net (US and Canada)
email: export@dynamax.com   (international)
http://www.dynamax.com

Easy Test, Ltd, Solarza 8b, P. O. Box 24, 20-815 Lublin 56
POLAND
Contact: Marek Malicki
(81) 744 5061, (81) 744 5067 FAX
email: henryk@gaja.ipan.lublin.pl

Edgcumbe Instruments Limited, Main St., Bothwell,
Glasgow G71 8EZ SCOTLAND
Contact: Jim McRae
(44) 1698 852574, (44) 1698 854442 FAX
email: support@edgcumbe.co.uk
http://www.edgcumbe.co.uk

EG&G MSI, 2450 Alamo Avenue SE, Albuquerque, NM 87106
(505) 243-2233, (505) 243-1021 FAX
http://www.egginc.com

Electro Rent Corp, 6060 Sepulveda Blvd.
Van Nuys, CA 91411-2512
(800) 688-1111, (818) 787-2100, (818) 786-4354 FAX
http://www.electrorent.com

Environmental Sensors, Inc., 14257 Barrymore St.
P.O. Box 720698, San Diego, CA 92172-0698
Contact: John Johnston
(800) 533-3818 (North America), (619) 484-0147 (International)
(619) 484-7212 FAX
email: jjohnsto@envsens.com
http://www.envsens.com

ESI Environmental Sensors, Inc., 100 - 4243 Glanford Avenue,
Victoria, B.C., V8Z 4B9 CANADA
Contact: Michael Marek (North American Inquiries)
Pierre Ballester (International Inquiries)
(800) 799-6324 (US and Canada), (250) 479-6588,
(250) 479-1412 FAX
email: mmarek@esica.com
email: pballester@esica.com
http://www.esica.com

GE Capital TMS, 6875 Jimmy Carter Blvd., Suite 3200,
Norcross, GA 30071
(800) GE RENTS
http://www.ge.com/capital/techmanage

GeoTDR, Inc., 297 Pinewood Drive, Apple Valley, MN 55124
Contact: Kevin O'Connor
(612) 431-3415, (612) 892-0241 FAX
email: kevin@geotdr.com
http://www.geotdr.com

Hewlett Packard, P.O. Box 58059, MS 51LSJ, Santa Clara,
CA 95052-8059
Contact: Test and Measurement Products
(800) 452-4844
email: RICHARD_GEORGE@hp-usa-om7.om.hp.com
http://www.hp.com

HYPERLABS, 13830 S.W. Rawhide Ct., Beaverton, OR 97008
Contact: Agoston Agoston
(800) 354-9432, (503) 524-7771, (503) 524-6372 FAX
email: agoston@hyperlabsinc.com
http://www.hyperlabsinc.com

IMKO GmBH, Im Stoeck 2, D-76275 Ettlingen GERMANY
Contact: Robin Fundinger
(49) 7243-592110, (49) 7243-90856 FAX
email: info@imko.de
http://www.imko.de

Irricrop Technologies Pty. Ltd., PO Box 487, Narrabri, NSW
AUSTRALIA 2390
Contact: Stephen Laird
(61) 67 922588, (61) 67 923804 FAX
email: sales@irricrop.com.au
http://www.irricrop.com.au

KANE GeoTech, Inc., P.O. Box 7526, Stockton, CA 95267
Contact: Bill Kane
(209)472-1822, (209)472-1822 FAX
email: wkane@kanegeotech.com
http://kanegeotech.com

KROHNE America, Inc., 7 Dearborn Road, Peabody, MA 01960
(978) 535-6060, (800) FLOWING, (978) 535-1720 FAX
email: info@krohne.com
http://www.krohne.com

McGrath RenTelco, 1901 North Glenville Dr., Suite 401A,
Richardson, TX 75081
Contact: Bill Chapman
(800) 233-5807, (972) 234-2422, (972) 680-0070 FAX
email: billc@rentelco.com
http://www.rentelco.com/gp.htm

MESA Systems Co., 119 Herbert St., Framingham, MA 01702
Contact: John Kussman
(508) 820-1561, (508) 875-4143 FAX

PermAlert ESP, 7720 North Lehigh Avenue, Niles, IL 60714
Contact: Dmitry Silversteyn
(847) 966-2190, (847) 470-1204 FAX
http://www.permapipe.com

Picosecond Pulse Labs, Inc., P.O. Box 44, Boulder, CO 80306
Contact: James R. Andrews
(303) 443-1249, (303) 447-2236 FAX
email: info@picosecond.com
http://www.picosecond.com

Polar Instruments Ltd, Garenne Park, St. Sampsons,
Guernsey GY2 4AF UNITED KINGDOM
Contact: Martyn Gaudion
(44) 1481 253081, (44) 1481 252476
email: mail@polar.co.uk
http://www.polar.co.uk

Riser-Bond Instruments, 5101 N. 57th Street, Lincoln, NE 68507
Contact: Gregory D. Hackbart
(800) 688-8377
email: ghackbart@riserbond.com
http://www.riserbond.com

Signetel, Inc., 51501 NW Hayward Road, Manning, OR 97125
Contact: Peter C. LoCascio
(503) 324-6634, (503) 324-6634 FAX
email: peter@locascio.com

Soilmoisture Equipment Corp., P.O. Box 30025, Santa Barbara,
CA 93105
Contact: Herb Fancher, Sales Manager
(805) 964-3525, (805) 683-2189 FAX
email: sales@soilmoisture.com
http://www.soilmoisture.com

Tektronix, Inc., 100 SE Wilson Ave., Bend, OR 97702
Contact:
Rick Puckett
(541) 330-4447, (503) 627-8010 FAX
email: rick.t.puckett@tek.com
or
Curtis Smith
(541) 330-4427, (503) 627-8010 FAX
email: Curtis.L.Smith@tek.com
http://www.tek.com

TELOGY, 3885 Bohannon Drive, Menlo Park, CA 94025
Contact: Mali Dahl
(800) 835-6494, (415) 462-5376, (415) 592-2618 FAX
email: mdahl@telogy1.com
http://www.tecentral.com

Tenzor Co., Box 86, Gagarina Ave. 603009, N. Novgorod,
RUSSIA
Contact: Alexandr Andrianov
007 (8312) 335352, 007 (8312) 652220 FAX
email: andr@andr.kis.nnov.su

TestEquity Inc., 2450 Turquoise Circle, Thousand Oaks,
CA 91320
Contact: Peter Kesselman
(805) 498-9933, (805) 498-3733 FAX
email: sales@testequity.com
http://www.testequity.com

Troxler Electronic Laboratories, Inc., 3008 Cornwallis Road,
P.O. Box 12057, Research Triangle Park, NC 27709
Contact: Ron W. Phillips
(919) 549-8661, (919) 549-0761 FAX

Tru Cal International, Inc., 605G Country Club Drive,
Bensenville, IL 60106
Contact: Brian Lottich
(630) 238-8100, (630) 238-8101 FAX

University of Amsterdam, Laboratory of Physical Geography and
Soil Science, Nieuwe Prinsengracht 130, Amsterdam 1081 VZ
THE NETHERLANDS
Contact: Timo J. Heimovaara
(31) 20 525 7451 phone or FAX
email: th@fgb.frw.uva.nl

9 780367 399